DARWIN'S RACISTS

DARWIN'S RACISTS

Yesterday, Today and Tomorrow

Sharon Sebastian and Raymond G. Bohlin, Ph.D.

"Darwin's Racists: Yesterday, Today and Tomorrow," by Sharon Sebastian and Raymond G. Bohlin. ISBN 978-1-60264-393-2.

Published 2009 by Virtualbookworm.com Publishing Inc., P.O. Box 9949, College Station, TX 77842, US, Copyright © 2009 by Sharon Sebastian and Raymond G. Bohlin. All Rights Reserved. www.DarwinsRacists.com

This book was written to impact the public good and increase its knowledge base as educational source material.

The web addresses/URLs, quotes, articles and books referenced in both this book and footnotes are solely recommended as resources for readers for research. They do not imply in any way an endorsement by authors, publishers, editors or compilers, nor do they vouch for their content or availability for the life of this book.

Unless otherwise noted:

"Scripture taken from the HOLY BIBLE. NEW INTERNATIONAL VERSION. Copyright © 1973, 1978, 1984 International Bible Society. Used by permission of Zondervan Bible Publishers."

"Scripture taken from the NEW AMERICAN STANDARD BIBLE®, Copyright © 1960, 1962, 1963, 1968, 1971, 1971, 1972, 1975, 1977, 1995 by The Lockman Foundation. Used by permission."

Contact: VBW Publishing at: www.virtualbookworm.com.

Printed in the United States of America

ACKNOWLEDGMENTS

Everyone who reads this book will have a decision to make about themselves and the world in which they live. To enable the reader to do so, ideas from across a wide spectrum are presented. Innumerable accredited references and sources, more than are normally necessary or required, are made available for readers to use for research should they desire. As well, this book was also created as an educational text and for consideration for course study.

Though varied with opposing views, the referenced sources in this book in no way imply or represent endorsement of this book or its content. It is recommended that the reader independently seek out the source materials and references listed. Go to the websites, read the articles and buy the books and DVDs for as balanced, educational and informed view as possible.

Since words have meaning, which in turn produces thought, *Darwin's Racists* allows us to become educated about ourselves and about those with whom we share the planet. We wish to acknowledge the various views that may help the reader determine who they are, where they came from, and, most importantly, how they view the rest of us based on what they believe.

CONTENTS

INTRODUCTION

IMPACT ON THE HUMAN RACE

How does one human being justify what he or she does to another based on nationality, class, race or disability? What informs or instructs our behavior towards one another? Some people believe we are creatures of evolution. Others believe we are creations of God. Does it matter what our origins are? Does where we came from determine who we are, how others treat us or how we treat them?

Though science is proving us to be more alike than we realize, it is a fact that no one, regardless of skin color, escapes racism from others who appear different from what they are. In the year 2008, the United States of America elected its first black President. During the presidential primary elections, old attitudes from all quarters resurfaced, threatening again to leave an ugly stain on society. Today, as in years past, there are those who promote such dissentions in an attempt to fracture society and advance their own political and cultural agendas. Discounting such external agitators, the real question becomes: What is the root source of prejudice that is perpetrated by and exists against all people regardless of their ethnicity, nationality or skin color around the globe?

What will be explored in the following pages are the core causes of social, ethnic and racial strife that originated yesterday, exist today, and may be further imbedded into the social fabric of tomorrow. The following questions must be addressed. Does the origin of human beings, how we came into existence, or where we come from geographically—truly separate us into distinct categories? Is there something deep within the social fabric historically, scientifically or spiritually that supports that separation?

These prescient and timely questions, and more, will be explored as two conflicting and discordant worldviews are examined along with the history, theology and science that support them. Examined will be the differing impacts that creation vs. evolution have had on ethnic, racial and class distinctions in our society.

Does whether we believe in God as our Creator or in Darwin's Theory of Evolution affect how we treat each other? Some suggest that the same Darwinian mind-set that fueled Hitler's ovens is being taught as fact in schools today. If so, how might that impact the future of us all, both on an individual level and globally?

Most major religions profess God as Creator in opposition to Darwin's theory of evolution. Globally, there are people who hold a God-centered belief that we were created equal. Yet, others have accepted Darwin's hypothesis, along with his theory of evolution, that people with different skin colors evolve or advance at different rates as part of a natural process. With due consideration given those who find common ground between the two, the question truly becomes—do you believe in God or Darwin and why? Many artificially imposed ethnic and racial differences and boundaries have root in cultural dogmas that are in direct opposition to science and religion. The question of today and tomorrow is: Should humans remain separate based on limited differences, or is there substantial likeness among all people that unites us as one race—the human race?

The question of where humankind came from is the subject of great interest in today's public forum. Is your ancestral father the lowly ape? Or, were you created by the same God who created the universe? The answer—will have consequences of immense proportions for current and future generations. It affects everyone on a very personal level.

PART 1

YESTERDAY

Yesterday's mind-set often establishes today's attitudes about one another. What can we learn from the influences of the past? How important is the current impact on race and class discrimination by Darwin's Theory of Evolution on today's culture?

CHARLES DARWIN—THE MAN AT THE CENTER

A man named Charles Darwin has been at the center of the debate about human origin since he introduced his <u>theory</u> mere generations ago in 1859. Seeking to please both his anti-Christian, pro-evolution grandfather and father, Darwin took up the family's banner of evolution. In his theory, Darwin formulated ideas about how the human race came into existence. Darwin hypothesized that all life on earth, including humans, is related and that everyone and everything—plants, animals, the birds in the air—evolved from a common ancestor. Darwin called his theory the "Theory of Evolution".

Darwin's theory of evolution had a major impact on how humans think of themselves and the world around them. In particular, in his second book, *Descent of Man,* Darwin stated in his own words that he wanted to assess the value of racial differences. In the introduction, he described his purpose for writing the book:

> *The sole object of this work is to consider, firstly, whether man, like every other species, is descended from some pre-existing form; secondly, the manner of his development; and thirdly, the value of the*

differences between the so-called races of man.[1]
Charles Darwin

Though his theory of evolution was not supported by scientific evidence, Darwin used it to analyze racial differences. He theorized that, by natural selection, one race might be superior to another based on outward physical appearances. In his book, *Descent of Man*, Charles Darwin further concluded that the human species evolved from animals and that the existence of different races of humans was evidence of a breakdown into sub-species or variants. In his own words, chapter one - page one, of *Descent of Man*, Darwin explained it this way:

> *It might also naturally be enquired whether man, like so many other animals, has given rise to varieties and sub-races, differing but slightly from each other, or to races differing so much that they must be classed as doubtful species?*[2] **Charles Darwin**

Darwin boldly wrote that man was a product of nature. In his book, *On The Origin Of Species*, first released in 1859, he not only presented his theory of evolution, but in effect, set the stage for those who rallied to both support his theory and nullify the belief in God as Creator. Darwin's theories stirred such heated controversy during his time that it has lasted well into today.

CHARLES DARWIN [3]
1854- 5 years before writing *On The Origin of Species*

Clearly, such heated debate demands a closer look. A closer look at how the theory of evolution took root and a review of what has carried it along until today. Again, what has been evolution's impact? Has Darwin's theory advanced the world community or added to its suffering? Is Darwin's theory of evolution racist at its core or are we truly evolving at different rates from one another? It should be noted that Darwin's book, *On The Origin Of Species By Means of Natural Selection*, carried the full title, *On The Origin Of Species By Means of Natural Selection or The Preservation of Favoured Races in the Struggle for Life.*

What exactly did Darwin mean by...*The Preservation of Favoured Races in the Struggle for Life*? Who did Darwin consider to be the favored races? If this is Darwin's foundation for his theory of evolution, that some races are favored and others are not, does such a theory sow the seeds of racism and class discrimination? If so, why is Darwin's theory being taught today in public schools to children of all ages?

In *The Descent of Man*, Darwin wrote these words as he theorized about human beings:

...the civilized races of man will almost certainly exterminate, and replace the savage races throughout the world.[4] **Charles Darwin**

Darwin further wrote...

The break between man and his nearest allies will then be wider, for it will intervene between man in a more civilized state—as we may hope than the Caucasian and some ape as low as a baboon— instead of as present—between the negro or Australian [Aborigine] *and the gorilla.*[5] **Charles Darwin**

On what basis did Darwin reach his conclusions in the mid-1800s? With no valid scientific evidence and no existing fossils of his supposed ape-like human ancestor, Darwin drew from his own speculations and prejudices. Based primarily on the appearances of various peoples of color, Darwin wrote his theory of evolution. Darwin took it upon himself to determine the differences or value of fellow human beings.

With Darwin came a pivotal change regarding the origins of humankind that pitted God as Creator of all versus Darwin's Theory of Evolution. It is of great importance to understand the impact of creation versus evolution in our lives.

CHAPTER 1

EVOLUTION

E volution. What is it? What impact does it have on human beings, on life itself? Evolution, to date, remains a theory on which scientists across the globe cannot agree. Born out of pagan belief, most devout evolutionists believe that humankind began as an "accident" of nature. That is in direct opposition to most of the world's major religions, including the Christian, Jewish, and Muslim faiths, that do not agree that life began as an accident but point to God as the Creator of the universe and all life within it. In a debate at the Cato Institute in Washington, D.C., Dr. Jonathan Wells, a renowned scientist in Molecular and Cell Biology from the University of California at Berkeley, said:

> Evolution can mean simply "change over time" or "change within existing species"—neither of which is the least bit controversial.[1]

It is true that all organisms, including humans, have the ability to adapt to a variety of environments. But, adapting should not be mistaken for evolving. Dr. Jonathan Wells has also said,

> Evolution is a hypothesis still looking for evidence.[2]

Is that true? Is evolution still just a theory not yet supported by fact? It is important to understand that a theory is not a law; it is not a scientific fact. A growing list of science professionals at dissentfromdarwin.com, challenge Darwin's theory of evolution as a viable means to explain all of life as we know it.

Since 1859, when Charles Darwin published the first edition of his book, *On The Origin Of Species*, evolution began to seriously challenge the God of the Bible as being the Creator of all. In his book, Darwin argues against the creationist or God-centered view. Since the days of Darwin, the theory of evolution has become globally entrenched. The question becomes not just when evolution took hold, but how is evolution gradually supplanting the role of God in our lives today? How is it that the theory of evolution now plays a major role in defining the origins of humankind and subsequently, by its very definition, separates people across the earth into unequally evolved groups based on skin color? A critical question is: Is Darwin's theory of evolution racist at its core or is Darwin right that we are truly evolving at different rates from one another? The key to those answers is found in the question: Does Darwin's theory have any foundation in science?

The evolutionary worldview permeates today's culture, though many remain unaware of its influence. Evolution has long been described as a theory. A theory, by definition, is an unproved assumption. Therefore, scientists in laboratories must hold out hope or faith that some day they might be able to prove assumptions or theories. But, if that hope comes with an already entrenched or pre-established bias, that bias may blind them from seeking other answers or solutions. Even though scientific data does not support it, evolutionists today insist on calling the "theory" of evolution a fact even as that theory is continually being forced to morph or change as new data is uncovered. Much of the new data is revealing the theory of evolution to be invalid.

Inherent in the basic evolutionary belief is the notion that all life began as an accident. This challenges the belief in God as Creator, in effect saying that: God does not exist, or if He does exist, He plays no active role in the affairs of the universe or humankind.

The biology of evolution has long been defined as the change from something very rudimentary or simple to something more developed or complex. In other words, going from a lower form of existence to a more advanced form of existence. Have human beings evolved over time from a lower or simpler form to a higher and more complex state? Does science support that theory? Some say, no. Others say, yes.

There is a scientific process that no one can dispute. Whether life is human, plant, animal, or the earth itself— once in existence, all matter begins to die as its ability to exist decreases with time. This simple observation is an inevitable fact of life. The very definition of evolution, when applied to human origin and existing matter, contradicts what is known in science as the *second law of thermodynamics*. It is one of the basic laws of science itself and a physical law to which modern scientists adhere. It states that all matter naturally degenerates from a more ordered state to a less ordered state. In other words, all matter—human, plant or otherwise—goes from a better state to a worse state over time. It does not matter who you are or what you are, from the moment of existence the life-to-death process begins. As babies grow, they are also aging. It is reality and an accepted fact of life.

If evolutionists accept this basic scientific tenet, then what is this "theory of evolution" they so strongly advocate? It is worthwhile to look at what evolution means in the broader sense of human origins and the actual creation of life. Why have staunch evolutionists rejected God as Creator with this mere theory? It is important to understand why.

Oxford Professor and British zoologist, Richard Dawkins is considered one of evolution's most avid defenders. A professed atheist, Dawkins wrote in the *New York Times* the following often cited comment:

> "It is absolutely safe to say that if you meet somebody who claims to not believe in evolution, that person is ignorant, stupid or insane (or wicked, but I'd rather not print that)."[3]

Dawkins clearly did not hold back his derision of critics of evolution by calling people who do not believe in evolution "ignorant, stupid or insane". On the PBS documentary "In the Beginning: The Creationist Controversy" (May 1995), creationist Dr. Phillip E. Johnson explained such criticism this way:

> *Darwinian theory is the creation myth of our culture. It's the officially sponsored, government financed myth that the public is supposed to believe in, and that creates the evolutionary scientists as the priest-hood...So we have the priesthood of naturalism, which has great cultural authority, and of course has to protect its mystery that gives it that authority—that's why they're so vicious towards critics.[4]*

Dawkins wrote about the importance of Charles Darwin in regard to his atheism. Darwin essentially rejected the Creator God with his substituted theory of how life evolved on its own through natural processes. Richard Dawkins is often quoted as saying:

> *"Darwin made it possible to be an intellectually ful-filled atheist."[5]*

Dawkins is further often quoted as stating:

> *"Any creationist lawyer who got me on the stand could instantly win over the jury simply by asking me: 'Has your knowledge of evolution influenced you in the di-rection of becoming an atheist?' I would have to answer yes."[6]*

In support of the statement regarding evolution's impact on atheism, Cornell University professor William Provine wrote on the university's website:

> *"Evolution is the greatest engine of atheism ever invented."[7]*

Theologian Dr. John MacArthur, author and host of *Grace To You*, does not doubt that evolution's influence has drawn converts to atheism. MacArthur has written that:

> *"At this point, nothing has ravaged gospel preaching to the untaught world more than the history of evolution."*[8]

Avowed atheists appear to have their own religion—a belief in evolution. In a statement proffering that atheism is, indeed, itself a religion, Dr. Jonathan Sarfati, of Creation Ministries International based in Australia, has written of Richard Dawkins in regard to evolution and Dawkins' denial of God, that:

> *"...the Apostle of Atheism has a long way to go to make a convincing case for his faith."*[9]

Legendary atheist and British philosopher Antony Flew, who for the better part of his life agreed with Dawkins, reportedly reversed his views:

> *"I have been persuaded that it is simply out of the question that the first living matter evolved out of dead matter and then developed into an extraordinary creature.*[10]*"* **Antony Flew**, Emeritus Professor of Philosophy, Reading University, formerly one of the world's leading proponents of atheism.

Flew was reportedly much impressed with the MIT educated nuclear physicist, Gerald Schroeder's scientific discoveries on Genesis 1, which is the first book of the Bible:

> *That this biblical account* [Genesis 1] *might be scientifically accurate raises the possibility that it is revelation.*[11]

Certainly, all people who believe in evolution are not atheists. Many people of faith believe a Creator God has allowed His creations to change or adapt over time. Some even go so far as to say that God chose to allow life to come into existence through a naturalistic, unplanned process. Adaptation and change occur and will be addressed in later chapters. But, change and adaptation are just that, change and adaptation. They in no way can explain, nor can they replicate, the actual creation of life.

Naturalism and agnosticism.

Phillip E. Johnson, author of *Defeating Darwinism by Opening Minds*, is a graduate of Harvard University's School of Law. Stephen Goode of *Insight on the News* asked why is it many intellectuals are agnostic and prefer naturalism to the Creator God. In reply, Dr. Johnson drew from his own experience:

> *It follows along on my own experience of the intellectual arrogance that comes naturally to an academic winner, an academic gold medal winner such as myself. Scientific naturalism is a thing that's attractive to that sort of people because it says that the secular intellectuals are the people to whom the world should look for all wisdom.*
>
> *The secular intellectuals become the priesthood. Their cultural story dominates. It feeds their sense that they have a wisdom the masses don't have. Naturalism is their vehicle to replace the religious clergy with the scientific and intellectual professionals, the priesthood being the people who tell a society its creation story, and in this case the creation story being the naturalistic one. So, honestly, if you want to see real dogmatism unrestrained, you must go to the higher reaches of the academic world and the scientific profession because the natural checks on dogmatism aren't there.[12]*

Evolutionists have not been able to scientifically prove that nature created itself from nothing, though their efforts continue. Leading scientists, like I.L. Cohen, have long challenged their colleagues in the scientific community who cling to the theory of evolution.

> *It is not the duty of science to defend the theory of evolution, and stick by it to the bitter end, no matter which illogical and unsupported conclusions it offers. On the contrary, it is expected that scientists recognize the patently obvious impossibility of Darwin's pronouncements and predictions. Let's cut the umbilical cord that tied us down to Darwin for such a long time. It is choking us and holding us back.*[13] **I.L. Cohen, Darwin Was Wrong: A Study in Probabilities.**

Whereas many describe evolution as representing an elitist attitude, it was clearly the elite of Darwin's day that avidly supported and promoted the theory of evolution. Today, that theory continues to have much support worldwide despite the lack of conclusive scientific evidence necessary to turn it from a theory into a fact.

Today, just as it was in the mid-1800s, the theory of evolution remains a theory whose foundation is strongly based on belief. When it comes to race and class, does it matter if one believes in the religion of evolution or the religion of a Creator God? Those beliefs versus modern science will be explored in chapters to come.

CHAPTER 2

CREATION

W hat is meant by creation? To create is to bring something into existence. That sounds simple enough. But, when it comes to the universe and life itself, it is far from simple. Human beings did not create the vast universe they live in which includes the earth, our sun, moon and stars. Human beings merely procreate or reproduce life from existing matter. Most people across our world believe the original spark of life, the original creation of all life, came from God.

Most of the world's major religions believe in a Creator God, in particular the Christian, Jewish, and Muslim faiths. Those who believe in God as Creator find both scriptural and scientific evidence that life was no accident. Anyone who agrees that God brought the physical universe and humankind into existence is called a creationist. This could be as minimal as believing that God created the laws of physics, chemistry, mathematics, astronomy and even natural selection. Or, a creationist may believe that the universe and life came into existence over a period of six literal 24-hour days, 6,000 to 10,000 years ago. Regardless, both are creationists.

Creationism is simply a belief in a higher power outside our known universe that gave structure and form to all that exists. Literally thousands of scientists around the world look to a Creator God as the explanation for the existence of the universe and all life. More scientists are increasingly skeptical that pure naturalistic or evolutionary forces provide an explanation.

For thousands of years people all over the world have accepted that a Supreme Being beyond themselves brought our world into existence. Today, things are not much different. It is estimated that there are over 2 billion people who would claim to be Christian; another 1 billion would claim to be Muslim. Both of these religions are described as theistic. That is, they believe in a transcendent God who created all things. Another 40 percent of peoples around the world still practice some form of a folk religion often referred to as animistic—meaning they believe in a form of life force throughout the physical realm that can be expressed through spirits, lesser gods and even the spirits of ancestors. But even these folk religions accept some kind of distant yet powerful Creator God who is over all.

The idea that God is some distant deity, who is not involved in the affairs of humankind or nature, is fairly recent in human history. This is usually referred to as deism. Even more recent, however, is the notion that there is no God at all. That is called atheism. And then there is the agnostic who just isn't sure.

In March of 2009, according to the American Religious Identification Survey, which reportedly queried some 54,000 people, more than two-thirds of adult Americans believe in God.[1] Among those remaining, some people are not sure. And, though their numbers have increased by a small percentage, atheists or agnostics still remain a small minority in the United States. It has long been noted that the largest percentages of nonbelievers come from the scientific community, though some of the most renowned scientists, whose work is revered and studied today, found that their personal belief in God strongly motivated them in their work and research.

> *A bit of science distances one from God, but much science nears one to him.*[2] **Louis Pasteur**, French scientist, developer of the pasteurization process.

A brilliant scientist held in great esteem for his contributions to mankind, Pasteur, also said:

> *...the more I study nature, the more I stand amazed at the work of the Creator.*[3]

Unlike Pasteur, many secular scientists assume that they have a greater knowledge and understanding, thereby making God superfluous. Creation scientists across the globe say that modern science has disproved Darwin's theory of racial inequality and, as a result of the lack of evidence to support it, that evolution is bankrupt as a scientific theory.

Yet, racism remains a moral issue that confronts the scientific community. Many Nazi scientists were racists against Jewish people and people of color. Their findings were once upheld as the science of the day in some quarters. Should care be given to assigning authority to the opinion of the scientific community on matters of race? Should such an important moral issue use Darwin's theory of evolution as the standard? Pathlights.com states in Science vs. Evolution that: "Thinking scientists increasingly question such an obsolete theory", as it quotes James Gorman from *The Tortoise or the Hare?*

> *Evolution...is not only under attack by fundamentalist Christians, but is also being questioned by reputable scientists. Among paleontologists, scientists who study the fossil record, there is growing dissent from the prevailing view of Darwinism.*[4] **James Gorman**

History records how man treats man based on Charles Darwin's theory of evolution. Some world leaders have used it to give rise to man's inhumanity to man in ways that Darwin may never have envisioned. If the theory of evolution breeds such inhumanity, what has been the impact of Creationism on the world community?

What does the word of God say about who we are and how we should treat one another? Who are we in the eyes of God according to the Bible?

BIBLE:

Then God said, "Let us make man in Our image, in Our likeness, and let them rule over the fish of the sea and the birds of the air, over the livestock, over all the earth, and over all the creatures that move along the ground." Genesis 1:26[5]

Note that the Creator said that humans are to rule over the earth and its plants and animals. Nowhere did God say that any one group of men or women is superior—or was given dominion to rule over any other group outside of the authority of God.

*"...male and female He **created** them." Genesis 1:27*[6]

"...the Lord God formed man from the dust of the ground, and breathed into his nostrils the breath of life; and man became a living being." Genesis 2:7[7]

The Bible clearly states that <u>God</u> created humankind. Secular scientists have been unable to disprove God's hand in creation, though they have unceasingly tried. All forms of reproductive life whether plant, animal, or human are reliant on existing living cells to procreate. The most advanced scientists of today do not begin to pretend to know how to create a human being from the dust of the earth, much less out of nothing. Scientific research is reliant on *existing living cells* to initiate cloning and genetic experimentation.

Did God create humans from the "dust" of the earth as the Bible says? In 1982, renowned scientist and Emeritus Professor of Cell Biology at the University of London, Dr. Edmund J. Ambrose, wrote the following in his book, *The Nature and Origin of the Biological World*:

At the present stage of geological research, we have to admit that there is nothing in the geological records that runs contrary to the view of conservative creationists, that God created each species separately, presumably from the dust of the earth.[8] **Dr. Edmund J. Ambrose**

To date, modern science supports what Creationist scientists have said all along...that the origination of the initial "spark of life" does not exist and has never occurred outside the realm of the Creator God. Neither humans nor science can create life, they can only replicate it. According to the word of God, we were all created the same in the sense that we all have a body, soul and spirit. The Bible proclaims that we all share the same heritage.

BIBLE:

He made from one man every nation of mankind to live on all the face of the earth, having determined their appointed times and the boundaries of their habitation... Acts 17:26[9]

God proclaims that we are all of one blood. Though, dispersed by God to various parts of the earth, God, Himself, assures us that we are all of <u>one</u> race - the **human race**. It was Acts 17:26 of the Bible that moved British parliamentarian and born-again Christian, William Wilberforce, to bring about the end of the slave trade throughout the British Empire in 1807. Based on the Bible's teachings—that God made all nations one blood and that all men are created equal in the image of their Creator God, Wilberforce, joined by other Christians, spent his adult life attempting to free slaves from their human bondage. Wilberforce wrote in his diary:

God Almighty has set before me two great objects, the suppression of the Slave Trade and the Reformation of Society.[10] **William Wilberforce**, October, 28, 1787.

Driven by his Christian beliefs in God as Creator, Wilberforce's actions led to the eventual abolition of the slave trade, not only in Great Britain and its colonies, but in the United States as well.[11]

Simply put, *Creationists* recognize <u>God</u> as *Creator*. After years of indoctrination by Darwinism, in 1929, British zoologist and evolutionist, D.M.S. Watson, said:

> *The theory of evolution is universally accepted by zoologists, not because it has been observed to occur...or can be proved by logical coherent evidence, but because the only alternative, special creation, is clearly incredible.*[12]

Creation is, indeed, incredible. According to the Bible...which is the best selling and most read book in history...the biblical God is sovereign over everything.

CHAPTER 3

CHARLES DARWIN'S FAMILY & FRIENDS

Again, the question is: How does one human being justify what he or she does to another based on race or class? What instructs or informs our behavior towards one another? Does it matter where we came from or what our origins are? Many believe it does. As mentioned, Charles Darwin has been at the center of the debate since he introduced his theory mere generations ago.

Who was this man Darwin—a man of privilege, a British naturalist born in Shrewsbury, England in 1809? Why did many in the world embrace his theory of evolution as not only a means to delineate the races, but also as a means to deny a Creator God? Why did others use the evolutionary Ape Chart to give rise to Darwin's theory of evolution as justification to devalue and annihilate those they deemed lesser men and women of the human species? Does it matter that Darwin's theory is allowed to be taught in public schools and universities today? Since "racism" exists in all races, will Darwinism have any impact on how we treat one another in the future? Does it matter if we think we came from apes?

A popular complaint from non-scientists is that if we believe we came from an ape-like ancestor, then many people, particularly young adults are tempted to act like apes. While this may be an intentional exaggeration, there is a rather large grain of truth to it. The real crux of this concern is the devaluation of human life. If we as human beings truly are just another animal species, no more, no less, then what are we to make of ethics in general and our own personal morality specifically? The rules that societies have adopted then become nothing more than survival strategies that, according to evolutionists, evolved over millions of years.

According to this model, as new societal or environmental challenges arise, then the rules of society will need to change to ensure survival. Nothing is absolute: all is relative to the ultimate goal of survival and reproduction. When you add to this scenario that in a purely materialistic universe, any organism, including humankind, is just a complicated collection of molecules, then something else tags along. The thoughts in our brains would then be mere interactions of molecules between brain cell synapses. Therefore, no one thought is any more right or moral than another. Whatever works for you or me determines right and wrong regardless of what others think.

When this self-actualized "reality" is applied, people are truly free to act on whatever makes them happy at the moment regardless of what it is. A person can create meanings for his or her life, but it is a meaning that is an illusion. When we realize this, suddenly all bets are off. Some people will live their lives as if other people don't really matter except as a means to get whatever is needed at the moment. We would essentially behave as animals under those circumstances, a different sort of animal for sure, but an animal nonetheless.

Darwin foresaw the implications of his reasoning twenty years before the publication of *On The Origin of Species*. Desmond and Moore reflect on Darwin's thinking as contained in his notebooks in 1838:

> *If man was only a better sort of brute, where was his spiritual dignity, and if he self-evolved, what of his moral accountability to God. No more his Creator? Since moral accountability, with eternal punishment and rewards, was part of the fabric of society that too would crash.[1]* **Charles Darwin**

In his circles, Charles Darwin was very much a Victorian English gentleman. He was considered kind-hearted and compassionate, particularly towards his wife, Emma, and their children. He was, however, a man of great contradictions. While stating he was against slavery for instance, in his book *The Descent of Man,* he nevertheless conveyed the idea that some human lives were inferior or less valuable than others.

In chapter 6 of *On the Affinities and Genealogy of Man,* Darwin argues that he fully expected the civilized races of men to fully exterminate the savage races of men in just a few centuries. He also expected the anthropomorphous apes (gorillas and chimpanzees) to also become extinct. As a result, he believed that the gap between humans and animals would eventually be much greater than exists today. Darwin postulates that this higher form of man will come from the current Caucasian race. In his book, Darwin states that the current gap between apes and humans is between the gorilla on the ape side and the Negro or Australian aborigine on the human side.

> *The break will then be rendered wider, for it will intervene between man in a more civilized state, as we may hope, than the Caucasian, and some ape as low as a baboon, instead of as present between the negro or Australian and the gorilla.*[2] **Charles Darwin**

Darwin's foremost German disciple, Ernst Haeckel, made even more dramatic statements. According to Haeckel, if you want to draw a sharp boundary between the human races and the apes, "...you must draw it between the most highly developed civilized people on the one hand and the crudest primitive people on the other, and unite the latter with the apes". Elsewhere Haeckel identifies these cruder and primitive races as the Australian aborigines and the South African Bushmen, which he says, still live in herds, climb trees and eat fruit.[3] According to Haeckel, certain more primitive groups of "people" are more ape than human.

Darwin certainly did not invent racism. Prejudice because someone is "other" than us has probably always been a part of human existence. What Darwin did provide was a scientific rationale that justified racial prejudice. Implicit in Darwin's struggle for existence is that some forms of a species would be more fit for the current environment than others. From Darwin's vantage point, the Caucasian or European race was well underway to surpassing the other "human" races because of their intelligence, culture, and superiority in war as demonstrated routinely in conflicts between Europeans and any other race or culture to that point.

Evolution, however, was not Charles Darwin's original idea. For young Darwin, evolution was undeniably—a family affair. From an early age, Darwin was influenced by his grandfather, who died before he was born, and his father. Both were medical doctors and not scientists. Both were also openly anti-Christian with the grandfather worshiping the pantheistic god of nature. Darwin's grandfather, Erasmus, first introduced his ideas on evolution in his book Zoonomia in 1794. In his writings, *The Temple Of Nature*, he wrote:

> [Man] *should eye with tenderness all living forms, His brother-ants and sister-worms.*[4] **Erasmus Darwin**

Though his grandfather was neither a scientist nor a geologist,[5] Charles Darwin still adopted many of his grandfather's ideas no matter how far-fetched. In his journals, Darwin wrote that his grandfather linked the common human emotion of anger to insanity:

> *My grandfather thought the feeling of anger, which rises almost involuntarily when a person is tired—is akin to insanity. I know the feeling of depression also, and both these give comfort to the body.*[6] **Charles Darwin**

Charles Darwin's father, Robert, son of Erasmus, was a man of ambition who wanted his son to fall lockstep into his footsteps. When his son fell short of his expectations, the father expressed great frustration about young Charles' intellectual shortcomings and frequently denounced him. Discouraged by Charles' poor performance in school, Robert Darwin sent his son off to medical school where Charles quickly became bored and dropped out. Disappointed, his father exploded.

> *You care for nothing but shooting, dogs, and rat-catching, and you will be a disgrace to yourself and all your family.*[7] **Robert Darwin** (father of Charles)

Darwin was then shuffled off to Divinity School. He tried to convince himself that his was a true faith, but it proved only superficial as evidenced by his later actions. In 1838, Darwin purposely misled his 1st cousin and future wife, Emma Wedgwood. Spurred on by his vehemently anti-Christian father, Darwin concealed from the future Mrs. Darwin, a woman of faith—and her family—his increasing doubts about religion.

Yet, while questioning religion, Darwin was also having nagging doubts about his grandfather's book on evolution. Upon reading it a second time, he wrote...

> *"...I was **much disappointed**; the proportion of **speculation** being so large to the facts given."*[8] **Charles Darwin** (emphasis added)

Once he married his cousin, the daughter of wealthy parents, Charles Darwin was determined not to live out his father's prediction that he would be "a disgrace to his family". Darwin took-up his family's banner of evolution. As a young man Darwin had left school with no science degree and no career on the horizon, though he had developed a fondness for beetle collecting and all things natural. Through the recommendation of one of his former professors, Darwin secured passage on the HMS Beagle for an around the world expedition as a naturalist and Captain's companion. Desiring to make his mark, on December 27, 1831, Darwin set out on a journey to parts of the world he had never seen before. Having limited knowledge of what had long existed in remote lands, he wandered into environments he found strange and wondrous in comparison to his native England.

Of the places that impacted Darwin most, were the Galapagos Islands and the island of Tierra del Fuego. To understand Darwin's theory of evolution, one has to ask: What happened on those islands that was so monumental that Darwin sought to reshape human beings' view of themselves and their origins? It is important to know how Darwinism began. It is important to revisit the birth of Darwinism—through Darwin's own words and observations.

In December of 1832, the ship he was on, the HMS Beagle, anchored off the Southern most extremity of South America known as the island of Tierra del Fuego. Darwin set off to explore the island. Using simple tools and often little more than a magnifying glass, Darwin would eventually make sweeping conclusions about animals, plants and native people he had never seen before. What electrified Darwin as new and strange was, in fact, normal and commonplace to the island people he would encounter. Acquainting himself with the native people, Darwin concluded the tribes' people of Tierra del Fuego to be "utterly strange savages" and sub-human.

These poor wretches were stunted in their growth, their hideous faces bedaubed with white paint, their skins filthy and greasy, their hair entangled, their voices discordant, their gestures violent and without dignity. Viewing such men, one can hardly make oneself believe they are fellow creatures and inhabitants of the same world. It is a common subject of conjecture what pleasure in life some of the lower animals can enjoy: how much more reasonably the same question may be asked with respect to these barbarians![9] **Charles Darwin**

Little doubt the islanders found the light skinned Europeans equally odd. Yet, it was Darwin who determined the dark skinned islanders to be a lower species of human. Darwin quickly surmised that the color of their skin and their habitat belied their potential for intellect. Ken Ham, of Answers In Genesis, says missionaries who lived among the tribes after the 1920s, did not reach the same conclusion as Charles Darwin.

> *After the 1920s, missionaries lived among the people of Tierra del Fuego. They found them to have a high standard of morality, believed in and prayed to a Supreme Creator Being, were generally kind and sociable, with respect for family life. A partial list of words in their language revealed they had a rich and complex grammar and vocabulary, which ran to more than 32,000 words.*[10]

Of the HMS Beagle's many ports of call that influenced Darwin's thinking on his expedition, no location was as significant as the Galapagos Islands, 600 miles west of Ecuador. The mere five weeks Darwin would spend there in the fall of 1835 would ultimately supply him with the necessary observations for his ideas of natural selection and what appeared to him to represent the common ancestry of most, if not all, of life.

His initial impressions of the islands as noted in his journal, *The Voyage of the Beagle*, were not too favorable, whether he was speaking of the climate or geography—"Nothing could be less inviting than the first appearance." Some of his other more colorful phrases regarding first impressions of plants and animals were "wretched-looking weeds", "dull-colored birds", "ugly yellowish-brown species" and "hideous-looking creature, of a dirty black colour, stupid, and sluggish in movements."[11]

Eventually he came to see that most plants and animals were "aboriginal creatures" found only on the Galapagos Islands, but with recognizable similarities to Central and South American species. These observations prompted him to write in his journal, *The Voyage of the Beagle*:

"Hence, both in space and time, we seem to be brought somewhat near to that great fact—mystery of mysteries—the first appearance of new beings on this earth."[12] **Charles Darwin**

New beings to whom? No doubt new to Darwin. Such a statement by Charles Darwin begs the question: In the short five weeks that Darwin spent on the islands, how was it he could conclude that new species had evolved and had become "new beings"? He certainly did not witness it. Most of the creatures across the world may have indeed appeared "new" to him, since he had never ventured far beyond the shores of his native England. Consider that the "hideous-looking creature of a dirty black colour, stupid, and sluggish in movements", that was so vile to Darwin, was merely a common island iguana.

Was Darwin jumping to extraordinary conclusions based on his previous lack of knowledge of other lands? That he was personally unaware that such animals and birds had existed for hundreds, if not thousands of years, makes them new only to Darwin, not new to the earth and certainly did not make them "new species" based on his sudden awareness of them.

Among the many specimens that he observed and collected on his journey were birds called finches. Darwin was not initially impressed by the finches he brought back to England. It wasn't until John Gould, an English ornithologist, had a look at them and announced that they were 13 different species, all quite similar and possibly from the same stock, that they gained notoriety.

Seeing this gradation and diversity of structure in one small, intimately related group of birds, one might really fancy that from an original paucity of birds in this archipelago, one species had been taken and modified for different ends.[13] **Charles Darwin**

Darwin paid little attention to this group of birds after his initial inclusion of them in his published journal, *Voyage of the Beagle.* He didn't even mention them in his landmark *Origin of Species* or *Descent of Man.* Some speculate that he hesitated to do so since he reportedly never bothered to keep track of which islands his specimens came from. Such reported inefficiencies in record keeping would surely cloud his later conclusions, which in effect could never be verified without making a return to the islands, which he never did.

But, many after him did return. David Lack spent six months in the Galapagos Islands and published his influential work on the Galapagos finches in 1940 and 1945. But, the latest research by Peter and Rosemary Grant, two British evolutionary biologists, began in 1973 and continued for three decades up to the present. The Grants visited the island of Daphne Major, in the Galapagos - every year to measure, weigh, count and record their findings. What we now have is a thirty-year record of change over time as recounted in their book, *Evolution in Action: Darwin's Finches of the Galápagos Islands,* which is available for reading. Jonathan Weiner in his Pulitzer Prize winning book, *The Beak of the Finch,* characterized their work with the highest superlatives.

> *This is one of the most intensive and valuable animal studies ever conducted in the wild: zoologists and evolutionists already regard it as a classic.*[14]

As Darwin suspected, the smallest of beak size variations in the ground finches could mean the difference between life and death. But, as others point out, this turned out to be a story of the bird's seasonal survival, not evolution.

The Galapagos experience yearly dry and wet seasons. The beginning of the wet season is the beginning of the breeding season. How long the wet season lasts influences how much breeding takes place. The Galapagos Islands, though right on the Equator, are a desert environment. The brief wet season, beginning in January, is a critical time for the finches. No rain—no seeds—no breeding season. Sometimes the wet season never comes for several years in succession. Then only the birds with the larger beaks survive since they can handle the larger seeds, which would normally be avoided by most of the other finches. The larger seeds are all that is left during these drought periods.

So, it at first appears the trend has different species of ground finches gaining in size, including both body and beak size. Indeed, natural selection seems to be favoring an increase in overall size across the species lines. But then, El Niño happens. The eastern Pacific warms by several degrees and torrential rains begin and last for months. Now all kinds of seeds are plentiful—smaller, softer seeds. Now to its detriment, the larger size is selected *against* by natural forces and smaller birds thrive. The size gains of the birds' beaks during the previous dry years are nearly wiped out in a single season.

Although he was only on the islands for a few weeks, Darwin's observations led to the speculation that "new species" had evolved on the islands. Clearly, he was not on the islands long enough to even witness the cycle that has been so well documented by the Grants for some thirty years. As noted, Darwin never returned to the islands after his initial five-week stop.

The Grants found potential for evolution in their tremendously detailed story, whereas, others view only adaptive changes. Even with beak changes or vacillations over time along with some habitat and behavioral variations, many scientists contend that the origin of a *new species* still seems nowhere to be found.

Certainly, many *variations* of finches exist. And, though many evolutionists may call them different species, is it a matter of semantics? Variations of finches are like variations of horses, whether it is a Palomino, a Pinto or Clydesdale, it is still a horse. Regardless of the variety of finches in many parts of the world or their oscillating beak sizes or varied behavior, a finch is still a finch whether it is the Galapagos Warbler Finch or the African Fire Finch.

What the evidence continues to show is an abundant demonstration of what has been called microevolution, which has never been in question. Microevolution is the variation within a species in response to environmental fluctuations. Again, what studies have proven is *adaptation* to environment, and not the evolution into a new life form.

So, while microevolution most certainly exists, there is no evidence of macroevolution or the *origin of new species.* There is no evidence that a new species has come into existence as the result of adaptive changes, small or significant. It is usually assumed that microevolutionary changes bring about macroevolutionary changes. But this has never been observed or documented by scientists or anyone else at any time in world history.

Regarding Darwin's finches, reportedly some school children still learn that during a severe drought the finches' beaks grew larger to adapt to a change in their environment. They are told that that adaptation is proof of evolution through natural selection since the finches' beaks grew larger to survive the drought. What needs to be taught is that the changes in the bird's beaks also reversed back to smaller sizes once the drought ended— with no apparent long-term trend towards evolution.

If Darwin were alive today, *he* would learn that the finches' beaks were merely oscillating back and forth during wet and dry years. It was merely a process of adaptation, not evolution, since the finches always remained what they always were and forever will be—finches.

When Darwin sent samples of the island's plant specimen home for analysis, he also indicated his limited understanding of the island's natural habitat. In a note he wrote:

> *I knew no more about the plants I had collected than the Man in the Moon.*[15] **Charles Darwin**

Though Darwin's grandfather heavily influenced his interest in evolution, it was Charles Darwin who introduced the idea of "natural selection". Natural selection being—that in nature, the weak eventually die out—while the strong that can adapt to their environment survive. Elitists of Darwin's day glommed onto the idea of natural selection. Helping the sick, sustaining the weak and aiding the poor—that occurs in all cultures, were no longer to be a part of the equation. In applying nature's process to man, compassion was to have no part in natural selection.

In 1864, after reading Darwin's exposes, British biologist, Herbert Spencer, was drawn to Darwin's concept of a lesser species of man. It was Spencer who introduced the phrase, "survival of the fittest" in his book *Principles of Biology*. Spencer is considered to be the father of Social Darwinism.

> This **'survival of the fittest'**, which I here sought to express in mechanical terms, is <u>that</u> which Mr. Darwin has called **'natural selection'**, or the preservation of the favoured races in the struggle for survival.[16]
> **Herbert Spencer** (emphasis added)

As one of Darwin's greatest admirers, Spencer's term "survival of the fittest" introduced a steroidal form of natural selection. With it came the introduction of Social Darwinism as Spencer also sought..."the preservation of the favoured races in the struggle for survival."

As Spencer misapplied Darwin's theories, many of the wealthy and elite of the day considered the poor and disenfranchised as being naturally selected by nature to die out. Subsequently, some elitists considered it an act against nature and a wasted effort to give assistance to the poor and disabled. It was clear to Darwin's elite that only the strong should survive.

Though Darwin held a more sympathetic view, his own words disclose his own thoughts that giving charity to the less able may prove to be a flawed practice:

With savages, the weak in body or mind are soon eliminated; and those that survive commonly exhibit a vigorous state of health. We civilised men, on the other hand, do our utmost to check the process of elimination; we build asylums for the imbecile, the maimed, and the sick; we institute poor-laws; and our medical men exert their utmost skill to save the life of every one to the last moment. There is reason to believe that vaccination has preserved thousands, who from a weak constitution would formerly have succumbed to small-pox. Thus the weak members of civilised societies propagate their kind. No one who has attended to the breeding of domestic animals will doubt that this must be highly injurious to the race of man. [17] **Charles Darwin**

Charles Darwin's brother, Erasmus Alvey Darwin, who was named after their grandfather, was quick to join his younger brother's supporters.[18]

In a letter to Charles, Erasmus Alvey Darwin wrote with scientific abandon:

"In fact the 'a priori' reasoning is so entirely satisfactory to me that if the facts won't fit in, why so much the worse for the facts is my feeling."[19]

Though Darwin's brother was gleeful at his elder sibling's newfound fame, there were others who foresaw ominous consequences. One of them was Darwin's close friend and former Professor Adam Sedgwick. After reading *On The Origin of Species*, he is said to have written:

"...I have read your book with more pain than pleasure. Parts of it I admired greatly; parts I laughed at till my sides were almost sore; other parts I read with absolute sorrow; because I think them utterly false and grievously mischievous. You have deserted—after a start in that tram-road of all solid physical truth—the true method of induction.[20]

Professor Sedgwick concluded his comments to Darwin with a warning:

> *"...if this book were to find general public acceptance, it would bring with it a brutalisation of the human race such as it had never seen before."*[21]

And truly, time showed that Sedgwick was right to have doubts. The twentieth century has gone down in history as a dark age when people underwent massacres simply because of their race, class or ethnic origins.[22]

A Godless Existence

Social Darwinism's influence is evident in its impact on race and class. But, some charge that inherent in the "evolutionary theory" is an atheistic denial of God. Many who flocked to Darwin's theories openly expressed their need to "do away with God." The intent being—if there is no God, man may do as he pleases.

The anti-God pro-evolution sentiment had Darwin championed at its core. During Darwin's day, among those desiring a Godless society was Thomas Huxley—both a good friend and dogged supporter of Darwin. Huxley's grandson, Aldous Huxley, openly wrote about his purpose for denying the existence of a moral God. In *Confessions of a Professed Atheist*, Huxley wrote that a moral God would hamper both his pursuit of power and his sexual proclivities:

> *I had motives for not wanting the world to have meaning—and consequently assumed it had none... The philosopher who finds no meaning in the world is concerned to prove there is no valid reason why he personally should not do as he wants to do, or why his friends should not seize political power and govern in the way that they find most advantageous to themselves.*[23] **Aldous Huxley**

Wanting a world without God and, therefore, without meaning, Aldous Huxley went on to clearly state that morals were an obstacle to his sexual desires:

> *The liberation we desired was simultaneously liberation from a certain political and economic system—and liberation from a certain morality. We objected to morality because it interfered with our sexual freedom.*[24] **Aldous Huxley**

No God meant no morality. Huxley's was a classic struggle of man against his Creator. Darwin's circle of supporters consisted of atheists, agnostics, and libertines who wanted an ever-present God out of their lives. They advocated the rights of the superior human over those they devalued as being of inferior class and race.

Darwin's Sickness

What of Darwin himself? What impact did his beliefs have on him as a person? On a personal level, for years as an adult, Darwin suffered prolonged and painful illnesses that his doctors were unable to diagnose. Among his illnesses were reportedly severe skin rashes, boils and a form of elephantitis that often left him disabled and unable to walk or work for vast periods of time. In a manuscript dated May 20, 1865, Darwin wrote to a medical examiner of his many afflictions:

> *For twenty-five years—extreme spasmodic daily and nightly flatulence: occasional prolonged vomiting. Vomiting preceded by shivering, hysterical crying, dying sensations or half-faint. Vomiting and every flatulence preceded by ringing of ears, treading on air, vision. Focus and black dots, air fatigues specially risky— brings on head symptoms and nervousness...*[25] **Charles Darwin**

Having reviewed Darwin's anguished writings in *The Life and Letters of Charles Darwin*, Dr. Edward J. Rempf wrote in 1918 on page 191 in the *Psychoanalytic Review*—that based on Darwin's writings, Darwin likely suffered from "anxiety neurosis".[26]

Some believe the causes for Darwin's ill health were both physical and psychological. In 1954, British psychiatrist, Dr. Rankine Good provided the following diagnoses:

> *"...if Darwin did not slay his father in the flesh, then he certainly slew the Heavenly Father in the realm of natural history suffering for his "unconscious patricide" which accounted for almost forty years of severe and crippling neurotic suffering."*[27] **Dr. Rankine Good**

Some speculate that Darwin was bitten by a black bug and contracted Chagas disease during his travels in South America. Medical analysis held such a theory suspect.

Dr. A. W. Woodruff, a British expert on tropical diseases pointed out that many of Darwin's symptoms (heart palpitations, undue fatigue and trembling fingers) appeared before Darwin sailed on the Beagle, and that when they recurred after his return, they were associated not with physical strain, but with "mental stress".[28]

That Darwin had his doubts and was stressed about his own theory is well documented by his own hand. Yet, those caught up in the opportunity to replace God with Darwin's theory of evolution, seized upon the chance. The new faith in human intelligence that would no longer require a higher source took root with a dogmatic determination. Darwin's theory began to replace religion in most laboratories and halls of higher learning. So writes physicist H. J. Lipton:

> *In fact, evolution became in a sense a scientific religion; almost all scientists accepted it and many are prepared to 'bend' their observations to fit in with it.*[29] **H.J. Lipton**

CHAPTER 4

EUGENICS' NIGHTMARE

O nce home from his travels and having published several books, including *On The Origin Of Species,* Darwin became somewhat of a reluctant celebrity. Though carrying on voluminous correspondence with scientists all over the world as a result of his published works, he declined most public appearances and would send his disciples to public meetings as his representatives and to give lectures. Most of these turned out to be developing atheists and agnostics.

While still under the powerful grip of his evolutionist father and grandfather, and his new friends, Darwin renounced the God of the Bible and decried the Christian faith.

> *I can hardly see how anyone ought to wish Christianity to be true, for if so, the plain language of the text seems to show that the men who do not believe, and this would include my father, my brother and almost all my best friends, will be everlastingly punished. This is a damnable doctrine.* **Charles Darwin**[1]

Among the friends and family of whom Darwin spoke was his cousin, Francis Galton, a man who embraced a form of early Social Darwinism that pitted society's "more fit" against its "less fit" poor and disabled. The devaluing of life based on race and class was taking root in the minds of Darwinists like Galton.

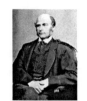

FRANCIS GALTON[2]

Seeking to take 'natural selection' to its extreme, Darwin's cousin, Galton, an atheist and racist, wrote in effect—if measures are not taken allowing inferior humans to survive and produce at levels faster than the more superior humans, society would be overwhelmed by inferiors. Galton wrote in his autobiography, *Memories of My Life*, that it was his cousin, Charles Darwin, who influenced his thinking on how to maintain racial superiority through selective breeding:

> *The publication in 1859 of the* On The Origin Of Species *by Charles Darwin made an epoch in my own mental development, as it did in that of human thought generally.*[3] **Francis Galton**

Based on his cousin Charles' theory of evolution, Francis Galton proposed the idea of instituting "breeding methods" for humans—much like those used for livestock and other animals. Galton coined the term "eugenics"—meaning "good birth". Through eugenics or "good birth", Galton hoped to prevent superior races from being polluted by what he considered inferior races.

Galton's eugenics was a rudimentary form of genetic engineering in that he wanted to use people with "better genes" to advance civilization. On the other hand, based on those with inferior genes, society could breed out of existence the lesser races, classes and the disabled.

To prevent co-mingling, Galton also proposed that marriages be legally regulated. He also believed regulations should be imposed regarding who qualified as genetically fit to have children. Galton's racist practice of eugenics fit perfectly with his cousin's newly introduced ideas of evolution and "natural selection".

Eugenics' goal would prevent those judged "unfit" from multiplying and, if necessary, be aborted. According to Galton, future generations depended on it. Years later, those who followed Galton's thinking established the British "Eugenics Society". No longer was evolution just affecting those of a different skin color. The Darwinist group proposed that all genetically weak, sick or disabled people should be sterilized—preventing them from reproducing. It was thought that in a "Eugenics Society" only the strong should reproduce.[4]

The problem for any "Eugenics Society" would be developing an objective criteria for who would be the "fit" and the "unfit". Even today, evolutionary theory only defines the "fit" after the fact since the fit are simply those who survive and reproduce in greater numbers. Predicting those who *will* be fit is a different matter altogether.

Such criteria end up being quite arbitrary and ill defined. Unfortunately, history has shown that the "unfit" invariably turn out to be the ones who least resemble those already in power. In 1911, succeeding his half-cousin Galton, the Eugenics Society's new chairman was Charles Darwin's son, Leonard [1911–1928].[5]

For Galton, if Darwin's theory could eliminate God, then man could do to another as he wishes without conscience or remorse. But eliminating God, especially in the late 19th century, proved not to be a popular concept. Yet, as others still continued to pursue his theory with vigor, Darwin opened the possibility that he was wrong, at least in his public communication.

> *If it could be demonstrated that any complex organ existed, which could not possibly have been formed by numerous, successive, slight modifications, my theory would absolutely break down.*[6] **Charles Darwin** (Origin of Species, 1859, p. 189.)

Regardless, Charles Darwin and his cousin, Francis Galton, held firm to the theories of evolution and eugenics...theories that soon began to etch a malignant mark on humanity. Over time, the outgrowth of eugenics proved horrific for the poor and the disenfranchised.

Once the idea became planted in the minds of racial elitists, its tentacles reached well into the twentieth century. In 1927, eugenics raised its ugly head in America. U.S. Supreme Court Justice Oliver Wendell Holmes, Jr., himself a self-professed Darwinist,[7] declared in a case before the U.S. Supreme Court that:

OLIVER WENDELL HOLMES, JR.

"Three generations of imbeciles are enough." **Oliver Wendell Holmes, Jr.**[8]

With that, the court allowed the forcible sterilization of tens of thousands of Americans that the state deemed "unfit". Thirty states passed compulsory sterilization laws in an attempt to stop reproduction of citizens believed to have defective genetic traits.

The support of Eugenics or "good births," which postulated that certain peoples with polluted human genes needed to be sterilized, was not limited to the United States. A number of nations used the practice worldwide.

Author Jonah Goldberg, Editor-at-Large of *National Review Online,* had the following to say in an interview with columnist John Hawkins:

Long before they (the Nazis) *started the campaign against the Jews, they started a massive euthanasia campaign, killing what they called the useless bread gobblers...What I think a lot of people don't appreciate and what has been pretty much well established now in the historical literature is that the Nazis were in so many ways picking up on ideas that first flourished in the United States under the progressives.*

The progressives start the forced sterilizations. It is the progressives who talk about weeding out the inferior races. Margaret Sanger, the Founder of Planned Parenthood, was all about weeding out the duskier and darker races...and all of that. The socialists of Britain, the Fabian Socialists, George Bernard Shaw, HG Wells, all of those guys, were soaked to the bone eugenicists who considered eugenics and socialism to be the same project. Oliver Wendell Holmes (U.S. Supreme Court Justice), *the author of the Bell Case, where the court ruled it was okay to sterilize low-income whites because they were viewed as sort of sub par genetic filth that needed to be cleaned up; Oliver Wendell Holmes said the first priority of social reform was to build a race. There were plenty of eugenicists who wanted to create a genetic Gulag Archipelago. They were going to put inferior women in, essentially, these interior colonies during their fertile years so that they could not breed.*[9]

After extensive research, Dr. George Grant, author of *Grand Illusions: The Legacy of Planned Parenthood*, concludes:

"Eugenics" referred to some people as "human weeds". Planned Parenthood's international work was originally housed in the offices of the Eugenics Society and the two were intertwined for years.[10]

Though frequently ignored today, Planned Parenthood is considered a spin-off of the eugenics' movement in America.

MARGARET SANGER [11]

Margaret Sanger, born in 1879 in Corning, New York, was one of eleven children. A social Darwinist, she was hailed as the mother of the eugenics' movement in America. Sanger held unorthodox opinions about marriage and family values. In her monthly journal, *The Woman Rebel*, she wrote:

> *"The marriage bed is the most degenerative influence in the social order."*

Sanger further wrote:

> *"The most merciful thing that a large family does to one of its infant members is to kill it."*[12]

Such avant-garde statements left much of the public cold. As a result, Sanger set out to alter the public's disapproving view. Her first step to redeem her public image was to rename her organization and, thereby, eliminate the term eugenics and all of its negative connotations.

"Planned Parenthood" was a name that had been proposed from within the birth control movement since at least 1938. One of the arguments for the new name was that it connoted a positive program and conveyed a clean, wholesome, family-oriented image.[13]

Sanger's new image was to be in contrast to her 1922 book, *The Pivot of Civilization*, in which she wrote about her goals for eugenics and dealing with the 'genetically inferior races'. Author Edwin Black, *War Against The Weak: Eugenics and America's Campaign To Create A Master Race*, wrote:

> *But Sanger was an ardent, self-confessed eugenicist [an activist for a Master Race], and she would turn her otherwise noble birth control organizations into a tool for eugenics, which advocated for mass sterilization of so-called defectives, mass incarceration of the unfit and draconian immigration restrictions. Like other staunch eugenicists, Sanger vigorously opposed charitable efforts to uplift the downtrodden and deprived, and argued extensively that it was better that the cold and hungry be left without help, so that the eugenically superior strains could multiply without competition from the 'unfit'. She repeatedly referred to the lower classes and unfit as 'human waste' not worthy of assistance, and proudly quoted the extreme eugenic view that human 'weeds' should be 'exterminated'."[14]*

Sanger reportedly practiced "free love" while advocating restraint and abortion for others who were mostly poor, disabled or people of color.

Was Margaret Sanger one of Darwin's Racists? Was she truly an evolutionist? In 1920, on page 47, of *What Every Girl Should Know*, Sanger wrote:

> *The lower down in the scale of human development we go the less sexual control we find. It is said that the aboriginal Australian, the lowest known <u>species</u> of the human family, **just a step higher than the chimpanzee in brain development**, has so little sexual control that police authority alone prevents him from obtaining sexual satisfaction on the streets.[15]* **Margaret Sanger** (emphasis added)

Sanger then turned an eye toward population control. She also began to refer to most of America's new immigrant population as "human weeds".[16]

Immigrant flood—Puck magazine cartoon USA Oct. 3, 1888.[17]

The flood of poor European immigrants was not the only problem Sanger addressed. In the 1930s, Sanger reached out to Harlem—suggesting that the answer to both poverty and black families' "desperate situations" lay in "smaller numbers of blacks". Recognizing the influence black ministers wielded, she wrote in a letter:

> *We do not want word to go out that we want to exterminate the Negro Population, and the minister is the man who can straighten out that idea if it ever occurs to any of their more rebellious members.*[18]
> **Margaret Sanger**

In the 1940s, Sanger's followers had pushed what had become known as the Negro Project into the south. A supporter of Planned Parenthood, Dr. Dorothy Ferebee was quoted in a speech:

> *The future program should center around more education in the field through the work of the professional Negro worker, because those of us who believe that the benefits of Planned Parenthood as a vital key to the elimination of human waste must reach the entire population.*[19] **Dr. Dorothy Ferebee**

Eugenics In America

As previously noted, Eugenics or the preference for "good births" had already reared its head in London in the early 1900s. The Eugenics Society wanted all handicapped people to be sterilized. Charles Darwin's son, Leonard Darwin, as president of the Eugenics Society made the following address at the First International Eugenics Congress in 1912:

> *As an agency making for progress, conscious selection must replace the blind forces of natural selection; and men must utilize all the knowledge acquired by studying the process of evolution in the past in order to promote moral and physical progress in the future. The nation which first takes this great work thoroughly in hand will surely not only win in all matters of international competition, but will be given a place of honour in the history of the world.* **Leonard Darwin**, Presidential address, First International Eugenics Congress, 1912.[20]

Leonard Darwin remained president of the Eugenics Society for seventeen years.

Meanwhile, in the United States, evolutionists in the 1920s and 1930s pressed for sterilization laws that were eventually passed and enforced in some American states. It was from the United States that Germany first saw eugenics being applied.

As previously mentioned, abortionist Margaret Sanger was a leading advocate of eugenics in America. In her pursuit to make eugenics and sterilization a standard in America for people of color, Margaret Sanger expanded the movement to also target all people who were poor or disabled. According to Sanger, the wrong people should not be allowed to breed. The fear was that they would not only become a drain on society, but also pollute the gene pool.

Dr. George Grant notes Margaret Sanger's words from her book entitled *The Pivot of Civilization* as follows:

> *Throughout its 284 pages, Margaret unashamedly called for the elimination of "human weeds", for the cessation of charity, for the segregation of "morons, misfits, and the maladjusted" and for the sterilization of "genetically inferior races." Published today, such a book would be labeled immediately as abominably racist and totalitarian.*[21]

Dr. George Grant again reminds us, in his book on page 61, that Sanger became the so-called 'mother' of Planned Parenthood.

Was Margaret Sanger an advocate of Social Darwinism? Did she consider the poor and people of color a burden on society? Her words speak for themselves in a pamphlet she wrote while at the forefront of the eugenics movement in America:

> *It is a vicious cycle; ignorance breeds poverty and poverty breeds ignorance. There is only one cure for both, and that is to stop breeding these things. Stop bringing to birth children whose inheritance cannot be one of health or intelligence. Stop bringing into the world children whose parents cannot provide for them. Herein lies the key of civilization.*[22] **Margaret Sanger** (Margaret Sanger, "What Every Boy and Girl Should Know", 1915, pg. 140.)

From the early 1900s to the 1970s, thousands of people were being involuntarily sterilized. The eugenics movement reportedly claimed close to 65,000 victims in thirty-three American states. North Carolina was among the first to target both males and females categorized as poor, defective or disenfranchised. In the 1960s, among those thought to have deficient genes was a thirteen-year-old girl named Elaine Riddick. Poverty stricken, she lived with her grandmother on a small North Carolina farm. Riddick was featured in an April 23, 2005, *ABC World News Tonight* report by Keith Garvin.

Garvin reported that Riddick was assaulted and became pregnant at the age of thirteen. Social workers labeled her promiscuous and too feeble-minded to ever be a responsible parent. After giving birth in 1968, Riddick was sterilized without being told. She learned the truth years later when she married and tried to have more children.

> *"They took so much away from me," Riddick said, "they took away my spirit and my soul."*[23]

Elaine Riddick also told her story in a TV documentary, *Darwin's Deadly Legacy*, broadcast in 2007 by Coral Ridge Ministries Media, Inc. She further explained that the man that had raped her as a child of thirteen had threatened her with her life if she told anyone. Fearful, she obediently entered the hospital after her grandmother had signed papers to allow the state to handle the birth. The forms also permitted the state to perform the sterilization. It wasn't until she was in her twenties that her grandmother told her with great anguish that she did not understand what she had signed.

Though once desperately poor and judged to be illiterate and feeble-minded by the state, Elaine Riddick went on to earn a college degree. Tony Riddick, the son she had at fourteen, became an engineering consultant. Mr. Riddick explained why he and his mother chose to come forward:

> *We don't want history to repeat itself, so it is extremely important for all Americans that we not turn a blind eye when we see these things happening.*[24] **Tony Riddick**

Billionaire and TV impresario, Oprah Winfrey, was born in rural Mississippi in 1954. She has stated publicly that her early years were spent living in poverty on her grandmother's farm and that she was molested by male relatives. She has revealed that at age fourteen, she gave birth to a premature baby who died.

Winfrey's story closely resembles that of the North Carolina college graduate who was sterilized. The three strikes against the North Carolina grad were that she was poor, black and ruled to be mentally diminished, yet she went on to earn a university degree. Elaine Riddick and Oprah Winfrey appear to have remarkably similar backgrounds as young girls. Some escaped eugenics' arbitrary and damnable reach.

Where did such a horrific mind-set come from that allowed for the discarding of human beings? Dr. George Grant traces it back to Thomas Malthus, an English philosopher and economist from whom Darwin found inspiration. Malthus sought to control populations. Dr. Grant describes Malthus attitude about his fellow man as follows:

> *Malthus argued that it was a good thing to allow the poor and undesirable aspects of populations to essentially kill themselves off through ill health, and argued that charity and philanthropy to the poor was bad for the human race because it allowed for what he* (Malthus) *called the breeding of human weeds.*[25]
> **George Grant**

Clearly, the Eugenics movement did not consider that all of God's children are born equal. Flawed information in the hands of the wrong people—had devastating outcomes. Such behavior by governments, when based on Darwin's theory of evolution, is called Social Darwinism. As historically documented, the impact of Social Darwinism has been devastating to millions of innocent people. Many need only point to history as proof that existing in Darwin's theory are ingrained prejudices that have meant abuse, depravation and death for our fellow human beings.

Dr. Alveda King, a former professor and much sought after speaker and author, is the niece of the late Dr. Martin Luther King, Jr. Dr. King told CyberCastNews:

*Planned Parenthood is definitely a racist organization—
they have a racist agenda...Since 1970, there has been
something like 50 million abortions. About 17 million of
those have been blacks. It's black genocide. They are
killing our people and fooling us.*[26] **Dr. Alveda King**

As a child, Dr. King had firsthand knowledge of ingrained
prejudices. In her biography she recalls the following
during the years of the Civil Rights Movement, which was
led by her uncle, Dr. Martin Luther King, Jr.

*My family home was bombed in Birmingham, Alabama
in the heat of the struggle. Daddy's house was bombed,
then in Louisville, Kentucky, his church office was
bombed. I was also jailed during the open housing
movement.*[27] **Dr. Alveda King**

Through it all, Dr. King points to a systematic eradication
of the unborn as one of her foremost issues.

Had eugenics' advocates of the early to mid-1900s had
their way, there would have virtually been an endless list
of people whose parents were poor, disabled, or people of
color, who may have never been born.

Among those whose impact on culture and society would
have greatly been lost—would likely include the following:

1. **MARTIN LUTHER KING, JR.**- Civil Rights Leader;
   African American

*"I have a dream...that my four little children will one
day live in a nation where they will not be judged by
the color of their skin, but by the content of their
character! I have a dream today!"*[28]

2. **CLARENCE THOMAS**- Supreme Court Justice;
   African American

3. **CONDOLEEZA RICE**- former Secretary of State; African American

4. **ROD PAIGE**- former U.S. Secretary of Education; African American

5. **MOHAMMED ALI**- Heavy Weight Boxing Champion; African American

6. **HANK AARON**- Baseball Hall of Fame Player; African American

7. **WILLIE MAYS**- Baseball Hall of Famer Player; African American

8. **ROSA PARKS**- Mother of the Civil Rights Movement; African American

9. **STEVIE WONDER**- singer; African American; blind

10. **CESAR CHAVEZ**- rights activists; Hispanic; poverty

11. **ARTHUR ASHE**- Wimbledon Tennis Champion; African American

12. **SAMMY DAVIS, JR.**- singer, dancer, movie star; African American

13. **FAMILY OF MATT AND AMY ROLOFF**- activists for Little People; parents; dwarfs

14. **OPRAH WINFREY**- TV talk show host; African American

15. **JOHNNY CASH**- Country Western Legend;-Poverty

Johnny Cash was born on February 26, 1932 in Kingsland, Arkansas. He was born into a family of very poor sharecroppers and claims he almost died of starvation as a small child.[29]

Radio Talk Show Host and author, Reverend Jesse Lee Peterson, is Founder and President of BOND Action, Inc., in Los Angeles. Long an advocate for fathers taking responsibility for their children, Reverend Peterson states the following statistic:

> *Every day more than 1,500 black babies are murdered under the auspices of Planned Parenthood, which is government funded. This is a race issue.*
> **Reverend Jesse Lee Peterson**

Peterson is one of a group of pastors that claims that Planned Parenthood has perpetuated "genocide on the black community".[30]

> *"Sanger's plan has worked extraordinarily well over time—today numerous black religious leaders defend the 'right' of women to kill their unborn children. Since 1973, some 16 million black babies have been aborted. It's ironic to me that black leaders complain about racism, yet they promote one of the most racist practices in this country—abortion.*
>
> *Abortion has also given black men one more way to be irresponsible. Because of the weakness of the father and the lack of morality in the black community, many black women feel they have no other choice but abortion.*
>
> *I believe there is a reason for the silence of black churches and black politicians regarding abortion. They actually have much in common with the abortionists: Abortionists are simply using blacks for power and wealth, the same way much of the black clergy, black politicians and liberal elite whites have used blacks for years.*
>
> *It is time for America, but especially the black community, to come out of its state of denial and realize that true racism is the attack on the black unborn baby, started by Margaret Sanger and carried*

out by the liberal elite in this country. The solution to this problem is a strong belief in the Creator, strong families and self-respect. Most importantly, men must step to the forefront of this issue. They must return back to their proper state as men of character and as the head of their families, or the horrors we've already seen in this 'one nation under God' will be dwarfed by the horrors to come."[31] **Rev. Jesse Lee Peterson, Founder, BOND Action, Inc.**

Why have some people thought so little of their brothers and sisters of different colors? On what basis did Darwin's elites demean whole populations and relegate them to being inferior? Did Darwin inadvertently provide a scientific justification for their racist attitudes?

One of the spurious dogmas was that humans evolved from apes. Darwin's elites were quick to surmise that people groups living closest to ape habitats had somehow evolved out of the ape population. The people of Africa were denigrated as being on the lower rung of human advancement—a work in progress on the evolutionary scale.

Did Darwin's elite and the scientists of his day get it wrong? Did they base their conclusions on fact or opinion? Can it be then that humans **are not** a species in **transition**—in various stages from ape to man? If Darwin's followers were wrong, to what lengths would they go to defend his theory?

CHAPTER 5

DARWIN'S RACISTS

Though Darwin suffered great physical and mental anguish of unknown origin, it paled in comparison to what was to follow for millions of innocent people. Girded by his grandfather's unfounded speculations on the origin of man and his need to prove himself to a domineering father, Charles Darwin's theory challenged the God of the Universe as Creator. Though Darwin may not have intended—history records that his observations, as written in his books, *On The Origins Of Species* and *The Descent Of Man*—in time and in the wrong hands—became the harbingers of mass annihilations and racial discrimination.

Hoping he had found a "missing link" in what he called the "primitive races" in his evolutionary chain, Darwin set in motion a scientific foundation for racial superiority that exalted some men over others. His concept of the existence of a "sub-human" between civilized man and the gorilla became the mainstay of Social Darwinism.

In the hierarchy of American government, there is a belief in God as Creator. It is a belief reportedly held by most American Presidents. Were they right? What about other world leaders? Historically, among the world leaders who would reject God as Creator—were Karl Marx who is known as the father of Communism, Joseph Stalin, Adolf Hitler and others, who—under the guise of Darwin's theory of evolution—formulated racist policies that resulted in the bloodshed of millions. How did Marx, Stalin and Hitler, among others, become **Darwin's Racists?** Did they simply co-opt Darwin's theory as justification for instituting murderous regimes? Innocent lives hung in the balance.

History, itself, repeatedly records that Darwin's theory of evolution became the foundation of a living nightmare for millions. Who set out to exploit Darwin's theory and use it as a vehicle for their own agendas? Who became **Darwin's Racists?**

The ugly side of Darwin's legacy sprang up globally. Racists from every continent and countries which included, but were not limited to: Germany, Australia, Arabia, Japan, China, Africa, Russia and the United States used Darwin's theory of evolution to justify the mistreatment of those who were "different" or judged to be "lower" on the evolutionary scale.

History records the names of men who used Darwin's theory of evolution as a justification to slaughter millions. The most notable to date are:

Darwin's Racists:

Karl Marx (1818—1883): Father of Communism
Karl Marx was the father of Communism—an atheistic form of government that rejects God in all facets of life. Marxism substituted the "state or government" for God. His writings bred the Marxists movements and dictators that set out to annihilate those who stood in their way. Calling himself "a secret admirer" of Darwin, Karl Marx sent Darwin numerous gifts. Among them was Marx's own book, *Das Kapital*, in which he wrote:

> *"From a devoted admirer to Charles Darwin."*[1] **Karl Marx**

In a letter to Marx, dated October 1873, Darwin indicated that he had difficulty in following the book and did not fully read it.[2]

Vladimir Lenin (1870–1924): Union of Soviet Socialist Republics (Russia)

Marx' protégé in Russia was Vladimir Lenin—who learned well from the Darwin admirer. A ruthless leader, history records that his was a reign of terror awash in rivers of blood.

Leon Trotsky (1879–1940): Soviet Union (Russia)

Without God's laws to restrain him, Leon Trotsky said that Darwin's ideas *"intoxicated him"*, Trotsky wrote that:

> *"Darwin stood for me like a mighty doorkeeper at the entrance to the temple of the universe."*[3] **Leon Trotsky**

Trotsky used evolution as justification to pursue power in whatever means he wished. He fanatically persecuted the Christian church.

Joseph Stalin (1879–1953): Soviet Union (Russia)

Reportedly the Russian dictator Joseph Stalin became a devout atheist after reading Darwin and determined that evolution did not require conscience or morals. In the 1930s, Stalin initiated what became known as the Great Purge in the Soviet Union that consisted of political persecution and executions of millions.[4]

The following quote is often attributed to Joseph Stalin:

> *"A single death is a tragedy...a million deaths is a statistic."* **Joseph Stalin**

Free to torture and murder, Stalin became known in the history books as the world's worst mass murderer, responsible for killing over 20 million of his own people. Some estimates are as high as 40 to 50 million dead at the hands of Stalin and his bloody terror.

In 1983, Alexander I. Solzhenitsyn, winner of the 1970 Nobel Prize for Literature, gave an address in London in which he attempted to explain why so much evil had befallen his people:

...while I was still a child, I recall hearing a number of old people offer the following explanation for the great disasters that had befallen Russia: "Men have forgotten God; that's why all this has happened."

Since then I have spent well-nigh 50 years working on the history of our revolution; in the process I have read hundreds of books, collected hundreds of personal testimonies, and have already contributed eight volumes of my own toward the effort of clearing away the rubble left by that upheaval. But if I were asked today to formulate as concisely as possible the main cause of the ruinous revolution that swallowed up some 60 million of our people, I could not put it more accurately than to repeat: "Men have forgotten God; that's why all this has happened."[5] **Alexander Solzhenitsyn**

An article written in 1987, by educator Paul G. Humber recounts the testimony of Soviet intellectual Alexander Solzhenitsyn as to why Stalin was able to pursue his murderous rampage. Humber reports on Solzhenitsyn's connection of Soviet oppression and Stalin's terror to Darwin's theories on origin.

In his towering book, *The Gulag Archipelago*, Solzhenitsyn recounts an incident, which apparently took place in the mid-1930s at a district Party conference meeting in a Moscow Province. (The full incident is described in Solzhenitsyn's book.)

"The secretary (replacing an arrested one) *was paying tribute to Comrade Stalin. The group, including the new secretary, was standing and applauding their esteemed Leader. Even a single minute of feverish clapping consumes energy, but in this case it was important to sustain the "enthusiasm" much longer. Three, four, five minutes passed and more! Tired*

arms!—but who could risk stopping? Seven, eight, and nine minutes elapsed. It was absurd! Finally after eleven minutes (!), a local factory director stopped clapping and sat down. All followed suit, but that night the one who stopped first was arrested and given ten years! He was told, "Don't ever be the first to stop applauding!" Solzhenitsyn queries, "And just what are we supposed to do? How are we supposed to stop?" In harmony with the position of this article, he (Solzhenitsyn) *adds:*

Now that's what Darwin's natural selection is. And that's also how to grind people down with stupidity."[6]
Alexander Solzhenitsyn

Adolf Hitler (1889–1945): Germany
Adolf Hitler was an avid follower of Darwin. He wrote in his book, *Mein Kampf,* (meaning *"My Struggle"*) that he believed in "natural selection" and the "survival of the fittest" as propagated by Darwin. Hitler's reign resulted in the murder of some six million Jews as well as Slavs, Poles, Blacks, Gypsies, homosexuals, the physically and mentally disabled and other groups he deemed unfit to live.[7]

History reveals that Hitler had a plan. Hitler often referred to the creation of a Master Race. To reach his goal, Hitler sought to rid German society of those he considered inferior or undesirable.

In *Mein Kampf,* Adolf Hitler wrote:

"...the purity of the racial blood should be guarded, so that the best types of human beings may be preserved and that thus we should render possible a more noble evolution of humanity itself."[8] **Adolf Hitler** (emphasis added)

More than anything, Hitler feared the Jewish intellect. Author Jonah Goldberg notes that Hitler went to great lengths to discredit the whole populace:

> *Hitler has this long section in Mein Kampf where he concludes that Jews aren't human beings, that they're a different species.*[9]

Mao Zedong (1893–1976): China

With Darwin and Huxley listed as two of his favorite authors, millions died at the hands of China's Mao Zedong as he pursued his utopian Marxist-Communist dream.

Pol Pot (1925–1998): Cambodia

Inspired by communists Stalin and Mao before him, Cambodia's Pol Pot's bloodthirsty regime, fronted by the Khmer Rouge, committed mass genocide against his own people.

Saddam Hussein (1937–2006): Iraq

Saddam Hussein of Iraq sought to emulate Stalin and Hitler, who were both avid followers of Darwin's theory of evolution. With Communist mass murderer Stalin and Nazi mass murderer Hitler as his stated heroes, Hussein tortured, butchered and gassed to death hundreds of thousands of his own people throughout his bloody career.

Darwin's theory became the ideological basis for communism, fascism and Nazism. Author of *Darwin On Trial*, Phillip Johnson, and Professor Emeritus of Law at the University of California, Berkeley is quoted as saying:

> *The philosophy—evolution—that fueled German militarism and Hitlerism—is taught in every American public school, with no disagreement allowed.*[10] **Phillip E. Johnson**

Darwin may not have had the foresight to know that his theory, in the wrong hands, would give racial and class discrimination a false and deadly scientific respectability.

Evolving The Super Race

Perhaps a closer look at the deadly past of Darwin's theory of evolution can reveal to us some things about the potential future.

Some fifty years before the advent of Adolf Hitler, it was Germany that gave Darwin his first glimmer of hope in the late 1800s. Initially, after publishing his book, *On The Origin Of Species* in 1859, no one seemed to notice. Darwin feared his theories would quietly die away—until he heard of the stirrings in Germany where his idea of "superior races" struck a chord. It was there, that his theory began to take root. Darwin wrote with optimism in a letter to German Wilhelm Pryor in 1868:

> *The support which I receive from Germany is my chief ground for hoping that our views will ultimately prevail.*[11] **Charles Darwin**

During a period of hardship for the German populace, Darwin's "theories" prevailed in horrific ways not only on the Jewish people, but on people of color, the disabled and the disenfranchised as well. Targets of the Nazis' racial slaughter were primarily Jews, Poles, Slavs, Gypsies, the mentally and physically disabled and homosexuals—totaling some eleven million victims. Nazi Germany stands as an inhuman and destructive result of Darwinism in the hands of an evil elite. Richard Weikart, a Fellow with the Center for Science and Culture of the Discovery Institute is author of the book, *From Darwin To Hitler: Evolutionary Ethics, Eugenics, and Racism in Germany* ...

> *Among German historians, there's really not much debate about whether or not Hitler was a social Darwinist.* **Richard Weikart**[12]

Frederick Nietzsche[13]

One of Germany's most influential scholars was Frederick Nietzsche, 1844–1900, a devout atheist and avid supporter of Darwin's ideas. An alcoholic, Nietzsche had syphilis, had a sexual predilection for men and was known to say, "woman is a snake" and if you "domicile with a woman, bring a whip." After years of mental illness, he died in an insane asylum.[14]

Nietzsche also hated all religion—whether it was Christian, Jewish, Muslim, Hindu or Buddhism. In place of a moral God, he preferred a "master morality" to be established by a race of "ubermensch" or "supermen" who were to be followed and obeyed. Nietzsche's beliefs had no room for compassion for one's fellow man. For Nietzsche, Darwinism justified one race eliminating another for its own benefit. For him, human existence was all about survival as man need only answer to himself for his actions. Nietzsche boldly pronounced that, "God Is dead." After Nietzsche's death in 1900, comments soon followed that God pronounced... "Nietzsche is dead".

Plagued by economic upheaval and unrest after World War I, Germany, was an open wound. Such talk of "survival of the fittest" became a rallying cry for a downtrodden people. It was an ideology that, in time, and in the wrong hands, proved deadly. This was the setting that brought to the fore one of evolution's foremost racists.

It was Darwin and Nietzsche's ideas that greatly influenced a young up-and-coming politician in the German political strata—a man named Adolf Hitler.

Hitler [15] Darwin [16]

As a young man, Germany's future leader was greatly influenced by the growing evolutionist fervor among some of Germany's elites. As leader of the National Socialist German Workers Party, which became known as The Third Reich, Hitler's persecution and extermination of people who were not of pure Aryan descent is well documented. But, a fact little discussed is that Hitler was a staunch Darwinist...an avid believer in the theory of evolution. Adolf Hitler used Charles Darwin's theory of evolution as a blueprint for his bloody regime. As he co-opted Darwin's own words, Hitler justified the Nazi practice of "natural selection."

> *"...remembering that many more individuals are born than can possibly survive—that individuals having any advantage, however slight, over others—would have the best chance of surviving and procreating their own kind? ...any variation in the least degree injurious would be rigidly destroyed. This preservation of the favourable variations, I call Natural Selection."*[17]
> **Charles Darwin**, 1859.

In the above passage from Darwin, simply replace the word "individuals" with the word "people" and you can see how Darwin's words have been interpreted as directing harm to human beings. If those with absolute power determine who carried the variations that were "in the least degree injurious", then why wait for natural selection? By selectively eliminating these "injurious" individuals, as was Hitler's plan, he and others like him were just speeding up what nature would do on its own.

Was Hitler a Darwinist? As Professor of Modern European History at California State University Stanislaus, Richard Weikart writes in his book, *From Darwin to Hitler,* that:

> *Darwinism was a central aspect of Hitler's world-view. It drove pretty much everything he did.*[18]

Hitler used Darwin's view of "natural selection" to justify his racist, anti-Semitic and, eventually, anti-Christian suppression and brutality. Though there is no evidence that Darwin considered Jewish people inferior, Hitler used Darwin's theory of evolution as a means of justifying his attack on Germany's Jewish population. Hitler considered other races to truly be inferior, but he feared and felt threatened by the Jewish peoples' intellect and abilities. Hitler applied Darwinism as scientific justification in his quest to eliminate an entire population. For Hitler, the devaluing and elimination of human life—became a tool for the state. Sir Arthur Keith, one of the foremost anthropologists of the twentieth century stated that:

> *The German Fuhrer, as I have consistently maintained, is an evolutionist; he has sought to make the practice of Germany to conform to the theory of evolution.*[19] **Sir Arthur Keith**

Hitler aggressively pursued his campaign of "natural selection" through terror and bloodshed. In his well documented book, *From Darwin To Hitler*, Richard Weikart concludes his chapter on "Killing the Unfit" this way: "By reducing humans to mere animals, by stressing human inequality, and by viewing the death of many "unfit" organisms as a necessary—even progressive—natural phenomenon, Darwinism made the death of the 'inferior' seem inevitable and even beneficent."[20]

Unrelenting in his persecution, Hitler's Nazi propaganda continuously linked non-Aryans, in particular Jewish citizens, to "monkeys" in an attempt to dehumanize them. In reality, Hitler was threatened by Germany's Jewish citizens' ever-increasing advancements in economics and the arts and sciences—in relation to their small numbers in the population.

After Hitler began to eradicate those he feared, or thought inferior, he eventually set his sights on his nagging problem of Christianity. In 1937, Protestant churches issued a manifesto openly defying the Nazi regime's racist policies. Seven hundred pastors were arrested.[21]

Hitler is said to have proclaimed:

> *"One is either a Christian or a German, you can't be both."*[22]

After World War II, the Chief U.S. prosecutor at the Nuremberg Trials, Justice Robert Jackson, revealed that Hitler planned to arrest, assault and kill Christian pastors, take over the churches, and "re-direct" their congregations. Jackson quoted anti-Jewish propagandist and pornographer, Julius Streicher, who complained that:

> *Christian teachings have stood in the way of 'the racial solution' of the Jewish question in Europe.*[23]
> **Julius Streicher**

Documents on file at Cornell University, compiled by U.S. General William Donovan,[24] indicate that the Nazi regime perceived the doctrine of Jesus Christ as a serious threat. For the Nazis, the Christian doctrine, that all people are of one blood, was in direct opposition to their labeling some groups of people as inferior. For Hitler, Bible-teaching evangelical churches were undermining the Nazi regime's "belief" in a super-race. Hitler set out to dismantle the churches and expunge the Judeo-Christian doctrine of divine origin of humans from German society. Hitler's right hand man, top Nazi Martin Bormann proclaimed:

> *More and more people must be separated from their churches—and their organs...the pastors.*[25] **Martin Bormann**

By Design Or By Chance author, Denyse O'Leary, is a Canadian Science Journalist who was once an apologist for Darwin. After years of study, in a series of articles she says it is clear through historical data that Darwinism provided the Nazis with a justification for their actions for treating people like animals or lesser human beings.

> *Similarly, we are told—and are expected to believe—that Darwinism had nothing to do with the Holocaust. When I was a dhimmi for Darwin, I have said that myself at times, and thus escaped the charge of impolite extremism.*
>
> *And I now retract and will say what I know to be true, based on four and a half decades of reading and studying: Darwin was instrumental in discrediting the traditional way of looking at human beings. This is a fact that everyone admits and many celebrate. How often have you heard that Darwin's great achievement was to knock humanity off its pedestal and show that we are merely evolved animals, accidentally evolved at that? And that had everything to do with the Holocaust. O'Leary goes on to say, "Whether Jews were Orthodox, Reform, or atheists made no difference to them—because they [Nazis] treated humans as if they were animals."[26]*

As one who defended Darwin until years of research changed her mind, O'Leary now writes a deafening conclusion:

> *Darwin put racism on a supposedly scientific basis. In that respect, he enabled the most virulent racism of the twentieth century."[27]* **Denyse O'Leary**

Darwin, himself, may have been horrified by the claim that his theory of evolution helped lead to the moral climate that allowed the holocaust. The Nazi regime established criteria that it used to identify "insufficient" human traits. Based on those traits, the following are only a few of millions who may have been marked by means of

Nazi imposed "natural selection" to perish had they lived under Hitler's reign.

POSSIBLE TARGETS FOR ELIMINATION

1. **STEPHEN HAWKING**: internationally renowned astronomical physicist; physically disabled (ALS-Motor Neuron Disease)

   Stephen Hawking once said,

   "My hand may shake but my heart does not."[28]

   The heart was not a factor Hitler considered.

2. **ALBERT EINSTEIN**: atomic physicist; Jewish

3. **FRANKLIN D. ROOSEVELT**: U.S. President; physically disabled (Polio).

4. **QUINCY JONES**: musician/composer; African American

5. **GOLDA MEIR**: Prime Minister of Israel; Jewish

6. **RAY CHARLES**: singer/composer; African American

7. **LEONARD BERNSTEIN**: composer/musician; Jewish

8. **JESSE OWENS**: Olympic Champion; African American

9. **NELSON MANDELA**: President of South Africa; South African

10. **SIR JOHN GIELGUD**: award-winning actor; professed homosexual

11. **ITZHAK PERLMAN**: acclaimed violinist; Jewish; physically disabled (Polio)

12. **HELEN KELLER**: author; Disabled Rights Advocate; physically disabled (blind and deaf)

13. **BARNEY FRANK**: U.S. Congressman (Massachusetts); professed homosexual

14. **BARBARA STREISAND**: singer/director/actor; political activist; Jewish

15. **DAVID GEFFEN**: record and film magnate; professed homosexual

16. **BETTE MIDLER**: comedian/actor/singer; Jewish

17. **LOUIS FARAKAHN**: Nation of Islam; African American

18. **ELTON JOHN**: singer; professed homosexual

19. **SIR IAN MCKELLEN**: award-winning actor; professed homosexual

20. **JHAMAK GHIMIRE**: Nepal; renowned writer, Cerebral Palsy

21. **CHRISTY BROWN**: eminent Irish writer/poet; subject of Academy Award winning film *My Left Foot;* Cerebral Palsy

22. **ANTHONY ROMERO**: ACLU Executive Director; professed homosexual[29]

And what of Charles Darwin himself? Should his name be added to the list? Had Darwin lived in Germany under the Nazi regime, the irony is that Hitler would have likely targeted Charles Darwin for elimination. Darwin, himself, documented that his multiple illnesses often left him unable to walk, bedridden and incapacitated.

Hitler's super race had no room for the physically or mentally weak. Germany's Darwinian evolutionists touted a better society, a new world based on "survival of the fittest". If Darwin's own illnesses pitted him against "the fittest"—who were to be chosen to survive—Charles Darwin did not measure up.

As history records, German Philosopher Friedrich Nietzsche's racist rhetoric and writings were magnets to the Fuehrer Hitler and his henchmen who had goose-stepped their way through human suffering and blood to become Germany's elite. Nietzsche drew support from Darwin in his desire to stunt reproduction in those he believed inferior. Darwinism provided the basis for his idea of breeding humans that would allow him to more quickly develop his imagined "super race."

Regarding the history of Nazi Germany, less than a century ago, should one again consider the haunting statement by former UC Berkeley law professor Phillip Johnson?

> *The philosophy that fueled German militarism and Hitlerism is taught as fact in every American public school, with no disagreement allowed.*[30] **Phillip E. Johnson**

CHAPTER 6

FRAUDS, FAKES AND MISTAKES

A trocities worldwide have been committed as one so-called race or nation has attempted to cast itself as superior based on the unsupported theory that the human race is an evolving species. The speculation is that some races of people evolve or advance at a superior rate than others, implying that some less evolved are closer to the supposed ancestral ape.

It again begs the question: Can it be that humans are not a species in transition, evolving in various stages from ape to man? If those transitions actually occurred, where are all of the half-ape, half-men that should be walking the streets and working among us today? All that exists among us today are apes that are...well, apes...with no transitional evolution apparent. Did the evolutionary process simply stop for no reason or did it never begin?

Stating that, "God created the earth and everything in it, including the laws of science" has caused irrational reactions. It has caused some evolutionists to perpetuate great frauds upon both the public and the scientific community for many decades.

Armed with Darwin's speculations, evolutionists hoped to clearly demonstrate that "several missing links" between monkey and man must have existed. They sketched-out on paper their ideas of what may have occurred in their presumed monkey-to-man evolution over time.

When evidence could not change fiction to fact, several frauds were perpetrated. In some cases, analysis was merely a mistake. Then there are evolutionists who have contended that the transitions between monkey and man are simply extinct and that they have the fossils to prove it. However, they have yet to produce the evidence.

The late Dr. James Kennedy, host of the DVD, *Darwin's Deadly Legacy*, comments that:

> *...over 100 million fossils have been catalogued around the world. With no evidence in that vast fossil collection to support evolution, Darwin's theory has broken down.*[1]

Randall Niles of AllAboutTheJourney.org writes: The British Museum of Natural History boasts the largest collection of fossils in the world. Among the five respected museum officials, author Luther Sunderland interviewed Dr. Colin Patterson, Senior Paleontologist at the British Museum and editor of a prestigious scientific journal. Patterson, a well-known expert with intimate knowledge of the fossil record, was unable to give a single example of Macro-Evolutionary transition. In fact, Patterson wrote a book for the British Museum of Natural History entitled, *Evolution.*[2] When asked why he had not included a single photograph of a transitional fossil in his book, Patterson responded:

> *...I fully agree with your comments on the lack of direct illustration of evolutionary transitions in my book. If I knew of any, fossil or living, I would certainly have included them...I will lay it on the line—there is not one such fossil for which one could make a watertight argument.*[3]

If it is not in the fossil record, where is the evidence that Darwin was so counting on? Darwin's own words echoed concern that if it cannot be proven in the fossil record, his original theory of macro-evolutionary progression could not be substantiated.

One of Darwin's earliest supporters and a perpetrator of a major fraud that was exposed worldwide was Ernst Haeckel, a German philosopher and biologist. Haeckel made drawings of alleged animal embryos and a drawing of a human embryo and compared them side-by-side. He claimed his self-drawn animal embryos so resembled the human embryo that it was evidence that man and animals shared common ancestry. In his effort to further popularize Darwin, Haeckel said his drawings were conclusive proof of evolution. But, even in Haeckel's day, science did not exist in a vacuum.

At Jena, the university where he taught, Haeckel was charged with fraud by five professors and convicted by a university court. His deceit was exposed in *Haeckel's Frauds and Forgeries*, a 1915 book by J. Assmuth and Ernest R. Hull, who quoted nineteen leading authorities of the day.

> *"It clearly appears that Haeckel has in many cases freely invented embryos, or reproduced the illustrations given by others in a substantially changed form," said anatomist F. Keibel of Freiburg University. Haeckel's distorted drawings were labeled "a sin against scientific truthfulness."*[4]

In a 1997 interview in *The Times of London*, Dr. Michael Richardson, an embryologist at St. George's Hospital in London, stated:

> *This is one of the worst cases of scientific fraud. It's shocking to find that somebody [that] one thought was a great scientist was deliberately misleading. It makes me angry. What he [Haeckel] did was to take a human embryo and copy it, pretending that the salamander and the pig and all the others looked the same at the same stage of development. They don't...These are fakes.*[5]

Despite Haeckel being discredited as a fraud, his drawings still appear in many high school and college textbooks,[6] though some are now reportedly being removed. In his 2000 book, *Icons of Evolution*, Jonathan Wells documents the reaction of two evolutionists in finding this fraud exposed all over again. Douglas Futuyma, author of a leading college evolution text who used Haeckel's drawings, said he would review his issue for future editions. Though, he reportedly commented that..."the various embryos really are very similar—we are talking about pretty minor differences."[7]

Foremost evolutionist, the late Dr. Stephen Gould, could not have made it any clearer that the cycle must stop:

> *We do, I think, have the right to be both astonished and ashamed by the century of mindless recycling that has led to the persistence of these drawings in a large number, if not a majority, of modern textbooks.*[8]
> **Stephen Gould**

Many teachers are simply repeating what they themselves were taught in high school and college. Having no idea that Haeckel's embryos were found to be out-and-out frauds, bad information is being played forward to their students.

In addition to his discredited drawings, Haeckel was also known for his "biogenetic law", in which he suggested that the development of races paralleled the various stages of development of individuals. He advocated the idea that "primitive" races were in their infancies and needed the "supervision" and "protection" of more mature societies.[9]

Haeckel's biogenetic law primarily maintained that embryonic stages resembled the evolutionary history of previous forms in the ancestry of that organism. For example, the human embryo, and all other vertebrate embryos for that matter, resembled a fish since vertebrates supposedly have a fish as a common ancestor. This was thought to be powerful evidence for evolution and is the suspected reason for Haeckel's fraudulent embryo illustrations.[10]

Since Haeckel's own colleagues exposed him almost from the beginning, an important question needs to be asked. How did this fraud that immediately brought shame and ridicule on its perpetrator, still end up in many of our established evolutionary textbooks over a century later? Some claimed ignorance, which is disappointing at best and some, like Gould who admittedly knew about it for decades, seemed willing to let it pass until it was brought to light again. What other so-called evolutionary evidence might be getting an illegitimate free pass?

Piltdown Man and Friends

One of the longest lasting evolutionary hoaxes of all time was that of the highly publicized Piltdown Man. The so-called Piltdown Man's remains, a piece of a skull and a jawbone, were found in a gravel pit in Sussex, England, in 1912. For forty years, leading scientists believed that the Piltdown Man was their proof of a missing link.

Imagine their dismay and disappointment when it was eventually found out that a jawbone of an orangutan had been artificially connected to a human skull. Upon closer examination, it was found that the orangutan's teeth had been purposely filed down and its jaw chemically treated to make it appear fossilized and human-like. Why did it take forty years to reveal a blatant and not very sophisticated hoax? Evolutionists reportedly clung to the false evidence of the Piltdown forgery, simply wanting to believe.[11]

Gertrude Himmelfarb, in *The Darwinian Revolution* (1959), wrote that:

> *Piltdown Man was a disaster for evolutionary theory because so many scientists either welcomed it or rationalized it into harmony with their prejudices. However earnestly scientists may now dissociate themselves and their theory from Piltdown man, they cannot entirely wipe out the memory of forty*

*years of labor expended on a deliberate and not
particularly subtle fraud.*[12]

Still desperate to come up with the so-called "missing
links", others in the scientific community created more
"frauds" or "fakes", or wrongly interpreted skeletal
remains. As mentioned, Jonathan Wells is a biologist with
a Ph.D. from both Yale University and the University of
California at Berkeley and author of the book *Icons of
Evolution: Science or Myth*? He states that for decades,
students have been taught things about evolution that are
simply not true.[13] A gullible public was convinced that the
Piltdown Man was truly the missing link. Upon honest
examination, it proved to be a fake.

Piltdown Man- FRAUD—Total fake—the jawbone belonged
to a modern ape.

Another fraud, perpetrated in 1922, was the so-called
Nebraska Man. After a single tooth was found in Nebraska
in 1922, speculations grew as to its origins and claims
were made that the single tooth was sound scientific
evidence of a "missing link."

Stories were made-up and written about Nebraska Man's
supposed lifestyle and were touted as fact to the public.
When later examined by an anthropologist, it turned out
the much ballyhooed tooth belonged to a wild pig and was
not from "early man" or a missing link—as was alleged.

Still believing the hoax back in 1925, the pig's tooth was
used as "key evidence" at the infamous Scopes ("monkey")
Trial. A trial, in Tennessee, that gave great impetus to the
evolutionary cause. The Nebraska Man charade captured
the public's imagination as drawings were distributed
worldwide depicting what the imagined creature would
look like if it actually existed.

Nebraska Man- FRAUD—Whole story was created about
Nebraska Man's life based on the tooth of a wild pig.

Imagined "Nebraska Man." (1922) [14]

The artist, had he known, would have been accurate to draw not a man, not an ape, but a wild pig. The so-called Nebraska Man did not exist.

Java Man- NOT A MISSING LINK—The best determination so far—it was the skull of an ape—leg of a human/not a gibbon.[15]

(Source: Hank Hanegraaff, *The Face That Demonstrates The Farce Of Evolution*, Word Publishing, Nashville, 1998, pp.50–52).

The following depiction is the Frontispiece to Thomas Huxley's *Evidence as to Man's Place in Nature* (1863), comparing the skeletons of various apes to that of man. It was yet another effort by Huxley to support his friend Charles Darwin's theory of evolution.

Skeletons of the Gibbon Orang. Chimp. Gorilla Man [16]

Neanderthal- NOT A MISSING LINK—Neanderthal was declared to be a human species. Its skeletal remains reportedly had a stooped appearance indicating arthritis and rickets due to a lack of vitamin D. Archaeologists have concluded that Neanderthals were skilled hunters, "indicated a belief in an after-life and even practiced a form of Social Security for their aged and infirm." ...One skeleton that was found had a withered right arm that appeared to have been amputated above the elbow by a skilled surgeon.[17]

(Source: "Upgrading Neanderthal Man," Time Magazine, Science, Vol. 97, No. 20, May 17, 1971.)

Evolutionists contend that Neanderthals were not quite as human as modern humans. Research is being done on the Neanderthals' bone fragments to determine more about them.

Orce Man- NOT A MISSING LINK—Was found in Spain in 1982 and thought to be part-man. Its skull fragment was not human and reportedly came from a four-month-old donkey.[18]

Peking Man- NOT A MISSING LINK—Human bones found in the freezing glacial area of far northern China. In his book, *Bones of Contention*, Marvin Lubenow documents that many anthropologists admit there is little reason to even call it a separate species, the similarities to full *Homo sapiens* are so real.[19] Their small body structure is thought to be the result of a bad diet and the rigidly cold environment.

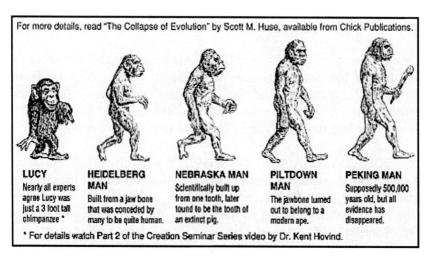

For more details, read "The Collapse of Evolution" by Scott M. Huse, available from Chick Publications.

LUCY
Nearly all experts agree Lucy was just a 3 foot tall chimpanzee *

HEIDELBERG MAN
Built from a jaw bone that was conceded by many to be quite human.

NEBRASKA MAN
Scientifically built up from one tooth, later found to be the tooth of an extinct pig.

PILTDOWN MAN
The jawbone turned out to belong to a modern ape.

PEKING MAN
Supposedly 500,000 years old, but all evidence has disappeared.

* For details watch Part 2 of the Creation Seminar Series video by Dr. Kent Hovind.

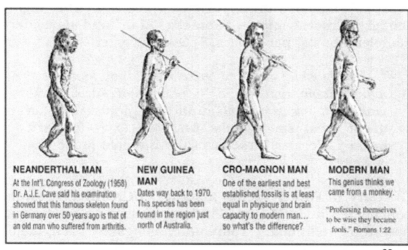

NEANDERTHAL MAN
At the Int'l. Congress of Zoology (1958) Dr. A.J.E. Cave said his examination showed that this famous skeleton found in Germany over 50 years ago is that of an old man who suffered from arthritis.

NEW GUINEA MAN
Dates way back to 1970. This species has been found in the region just north of Australia.

CRO-MAGNON MAN
One of the earliest and best established fossils is at least equal in physique and brain capacity to modern man... so what's the difference?

MODERN MAN
This genius thinks we came from a monkey.

"Professing themselves to be wise they became fools." Romans 1:22

20

The latest ape-to-man hope surrounds a collection of bones discovered in 1974 and named Lucy. Until recently, Lucy has been held up as the prime transitional fossil—said to represent an indisputable link between humans and apes. After a six-year tour of some of the world's top museums, Lucy has been found—<u>not</u> to be human at all. Anatomist David Menton says that Lucy is just one more so-called "human ancestor" that evolutionary science has been forced to <u>discard</u>.[21]

David Menton taught anatomy at Washington University School of Medicine and holds a Ph.D. in cell biology from Brown University. Menton is considered one of today's leaders in physical and biological science. One of Dr. Menton's lectures is titled "Lucy: She's No Lady".

In 2007, Tel Aviv University researchers find that Australopithecus or "Lucy," latest star of the evolutionists' stage show, is not a direct ancestor of humans. Lucy has the same wrist-locking mechanism of all <u>knuckle</u> <u>walkers</u>. Dr. Menton of Answers in Genesis says, no clear link exists between humans and apes after one hundred plus years of intense anthropological research. The highly touted Piltdown Man, Nebraska Man and Lucy are complete "busts" per Menton.[22] Others agree.

> *"...Lucy's skull was so incomplete that most of it was 'imagination made of plaster of Paris,' thus making it impossible to draw any firm conclusion about what species she belonged to."*[23] **Richard Leakey**, Former Director of the National Museums of Kenya.

Dr. Gary E. Parker, of the Institute of Creative Research, commented in 1981, that Piltdown Man is an example of how the evolutionary community clings to such major gaffes.

> *At least Piltdown Man answers one often-asked question: "Can virtually all scientists be wrong about such an important matter as human origins?" The answer, most emphatically is, "Yes, and it wouldn't be the first time." Over 500 doctoral dissertations were done on Piltdown, yet all this intense scrutiny failed to expose the fake. Students may rightly wonder what today's "facts of evolution" will turn out to be in another 40 years.*[24]

Darwin certainly expected transitions to be found with time. Since evolution is supposed to happen over time at a slow and gradual pace, we should today be able to witness its progressions all around us. The transitions should be everywhere and not very difficult to recognize or interpret. Yet, this is not the case.

Whether it would be the alleged ape to man transitions or those of animals, birds and even amphibians from fish, the fossil record does not support it. No one can elude the fact that if evolution were to have occurred, evolutionary transitions would be found all over the fossil record...and they are not.

Zoologist, David B. Kitts, as Head Curator of the Department of Geology at the Stoval Museum, contends that the fossil record in general has been a continuous frustration to evolutionists. In an evolutionary trade journal, he wrote:

> *Despite the bright promise that paleontology provides a means of "seeing" evolution, it has presented some nasty difficulties for evolutionists, the most notorious of which is the presence of 'gaps' in the fossil record. Evolution requires intermediate forms between species and paleontology does not provide them...*[25]

Consider Dr. Niles Eldredge, the staunch evolutionist who proclaimed himself a "true Darwinist". He states on his website that his "main professional passion is evolution."[26]

Dr. Eldredge is responsible for the Darwin exhibit at the American Museum of Natural History. Eldredge wrote in one of his books:

> *We have proffered a collective tacit acceptance of the story of gradual adaptive change, a story that strengthened and became even more entrenched as the synthesis took hold. We paleontologists have said that the history of life supports that interpretation, all the while really knowing that it does not.*[27]

Even in his day, Charles Darwin cited that the lack of evidence in the fossil record was "the most obvious and gravest objection which can be urged against my theory."[28]

Once more, add to that doubt Darwin's other realization...

> *If it could be demonstrated that any complex organ existed, which could not possibly have been formed by numerous, successive, slight modifications, my theory would absolutely break down.*[29] **Charles Darwin** (*On the Origin of Species by Means of Natural Selection, or the Preservation of Favoured Races in the Struggle for Life,* 1859.)

With that, Darwin himself disproves his theory of evolution based on his own stated criteria. Primarily drawing his conclusions from physical appearances, Darwin barely had a rudimentary understanding of the intricate inner workings of the human body. Biochemist Michael Behe, author of *Darwin's Black Box*, points out that Darwin's knowledge of a human cell was both simple and erroneous.

> *To Darwin, the cell was a 'black box'—its inner workings were utterly mysterious to him. Now, the black box has been opened up and we know how it works. Applying Darwin's test to the ultra-complex world of molecular machinery and cellular systems that have been discovered over the past 40 years, we can say that Darwin's theory has 'absolutely broken down'.*[30] **Michael Behe** (biochemist/author) *Darwin's Black Box* (1996)

Were he alive, Darwin would learn that another death knell for his theory is one of today's scientific actualities called "irreducible complexity". Briefly, irreducible complexity is the idea that some systems, like a modern mousetrap for example, are composed of two or more absolutely necessary parts. Take any one part away and the system ceases to function—at all. If you take the spring out of a mousetrap, you won't just catch fewer mice, you won't catch any mice.

Similarly in the cell, we have now identified hundreds of potentially irreducibly complex molecular machines that could not have been built gradually by Darwinian natural selection. All of the cell's necessary protein parts must be present at the same time. Gradual accumulation through the long evolutionary process yields functionless protein complexes that would be lost.

Molecular and Cell Biologist Dr. Jonathan Wells from the University of California at Berkeley has done a study of textbooks used in the public school system.[31]

> *Dogmatic Darwinists claim that nothing in biology makes sense except in the light of evolution. Then they misrepresent the evidence to promote their view. The truth is, nothing in biology makes sense except in the light of evidence.*
>
> *...Many leading high school and undergraduate biology textbooks present students with drawings of similarities between fish and human embryos, and claim that these similarities are evidence that fish and humans share a common ancestor. Embryologists have known for over a century that these drawings were faked.*[32] **Dr. Jonathan Wells**

What has been the impact of such fakery and fraud on how one person views another? Among the evolutionary beliefs that remain is that if humans evolved in Africa—Africans must be geologically closer to the origin of their ape ancestors and, by implication, must be biologically closer also. That assumption lingers despite the Bible and modern science relegating that hypothesis to mere <u>theory</u>.

Atrocities

The late Harvard Professor Dr. Stephen Gould, an avowed Darwinist, wrote:

> *Racists attitudes were common before Darwin, but racist attitudes greatly increased 'by orders of magnitude' following the accepted theory of Darwin's theory of evolution.*[33]

Considered to be the evolutionist's evolutionist in the twentieth century, Dr. Gould, a self-proclaimed atheist who devoted his life to studying evolution, went to his death—still unable to answer, how did life begin?[34]

As the esteemed Dr. Gould implied—do Darwinism and the theory of evolution fuel racism? One need only look to the "human zoos" that existed from the 1870s to just after World War II. Indigenous people from the Sudan, Samoa, to the Congo and other populations were on display from Paris to Berlin, from London to Barcelona, from Moscow to Warsaw and from St. Louis to New York. Many were viewed in settings described as Nubian villages.

> *Ethnographic zoos were often predicated on unilinealism, scientific racism, and a version of Social Darwinism. A number of them placed indigenous people (particularly Africans) in a continuum somewhere between the great apes and human beings of European descent. For this reason, ethnographic zoos have since been criticized as highly degrading and racist.* [35]

In 1904, an African tribal slave trader sold nine fellow Africans to be transported to the United States to be exhibited at the St. Louis World's Fair. Among them was Ota Benga, a healthy twenty-three year old male dwarf from the Belgian Congo in Africa, where people under five feet were commonly called pygmies. As one of the "natives" on display at the World's Fair, Benga reached the greatest notoriety when he was eventually exhibited in New York.

OTA BENGA—BRONX ZOO—1906 [36]

In 1906, Ota Benga was housed at the Bronx Zoo. For a while, he roamed the zoo freely and helped feed the animals. That was until Madison Grant, a prominent scientific racist and evolutionist encouraged the zoo director that there could be profit in exhibiting the young man in the Monkey House. Caged with an orangutan, Grant quickly labeled him the "missing link".[37]

Segments of the Christian public decried Ota Benga's treatment as being patently "racist". African American Clergyman James H. Gordon lodged a complaint saying Ota Benga's treatment was a blatant promotion of evolution.

> *Our race, we think, is depressed enough, without exhibiting one of us with the apes. The Darwinian theory is absolutely opposed to Christianity, and a public demonstration in its favor should not be permitted.*[38] **Reverend James H. Gordon**

When white clergy protested as well, The New York Times ran a front-page story on September 10, 1906, "Man and Monkey Show Disapproved by Clergy" which listed some of the complaints:

> *"The person responsible for this exhibition degrades himself as much as he does the African," said Rev. Dr. R. MacArthur of Calvary Baptist Church. "Instead of*

making a beast of this little fellow, he should be put in
school for the development of such powers as God
gave to him." The New York Times, 1906.[39]

Toward the end of September 1906, Ota Benga came
under the guardianship of Reverend Gordon, who placed
him in a church sponsored orphanage in Virginia of which
Gordon was the superintendent.

After briefly being tutored and sent to school, Ota Benga
dropped out and began working at a tobacco factory.
Missing his family and his homeland, on March 20, 1916,
at age thirty-two, Benga reportedly built a fire, performed
a final tribal dance and shot himself in the heart with a
stolen pistol.[40]

Aborigines and World Horrors

Among the worst documented incidents in history
regarding the treatment of the so-called "missing links"
occurred in Australia. Aborigines, a people indigenous to
Australia, are believed to have originally migrated from
Asia. In the twentieth century, their treatment as sub-
humans left an ugly scar on the land down under. In
1999, the government of Australia issued an open apology
to Aborigines essentially stating that it was sorry for the
stealing and killing and marked the occasion as the time
for the country to start healing. Most Aboriginals accepted
the heartfelt apology, but well know the effect of
Darwinism.

Ken Ham, Australian co-founder of Answers in Genesis,
relates the story of the horrors perpetrated on the
Aborigines.

Darwin taught (in his second book) *Descent of Man,*
that the Australian Aborigine was closet to ape-like
ancestors. The Australian Aborigines were looked
upon as missing links in history. In the early 1900s,
they were even listed as animals in a Sydney
museum. Scientists in Germany and England sent

researchers to Australia with instructions on how to skin the Aborigines, how to blow up their skulls for specimens for museums around the world. Aborigines were thought to still be evolving thanks to Darwin's theory.[41]

Clearly, a gruesome trade in 'missing link' specimens began with early evolutionary/racist ideas and rapidly spread with the advent of Darwinism. *Creation* magazine related evidence that perhaps 10,000 dead bodies of Australia's Aboriginal people were shipped to British museums in a frenzied attempt to prove the widespread belief that they were the 'missing link.'[42]

A major item in a leading Australian weekly, *The Bulletin,* revealed shocking new facts. Some of the points covered in the article, written by Australian journalist David Monaghan, were that:

Along with museum curators from around the world (Monaghan says), *some of the top names in British science were involved in this large-scale grave-robbing trade. These included anatomist Sir Richard Owen, anthropologist Sir Arthur Keith, and Charles Darwin himself. Darwin wrote asking for Tasmanian skulls when only four full-blooded Tasmanian Aborigines were left alive, provided his request would not 'upset' their feelings. Museums were not only interested in bones, but in fresh skins as well. These would provide interesting evolutionary displays when stuffed.*[43]

A German evolutionist, Amalie Dietrich (nicknamed the 'Angel of Black Death') came to Australia asking station owners for Aborigines to be shot for specimens, particularly skin for stuffing and mounting for her museum employers. Although evicted from at least one property, she shortly returned home with her specimens.[44]

Countries involved in such horrors were not limited to Germany, America, Russia, China, Cambodia, and

Australia. The tentacles of Darwinism reached around the globe. During World War II, Japan believed their Emperor was descended from a sun goddess and that the Japanese people were a highly evolved "master race". Mixing theism with Darwin's evolution, they pointed to the ape-ish (apish) body features of other races with their long arms, body hair, and pungent body odor.[45]

By the twentieth century, history notes that the Arab slave trade took a reported eighteen million slaves from Africa to the Muslim world where a bias remains today.

In Africa—from Botswana to Niger, to Mauritania and the Republic of Congo, atrocities have been committed as Africans shed their own blood over which tribal groups were more evolved than the other.[46]

> "How can we continue to have Stone Age creatures in an age of computers?" asked Botswana's President Festus Morgae.[47]

The United Nations' Committee on the Elimination of Racial Discrimination released a report that condemns Botswana's treatment of the 'Bushmen' as *racist*.[48]

Some call Darwin's speculation of superior or inferior status based purely on the color of a person's skin, both racist and dehumanizing. If they are right, what could that mean for society in the twenty-first century and beyond?

CHAPTER 7

THOUGHT POLICE

Today, as some people go about their lives, little do they realize that there is a serious debate about the origins of the human race. A great divide exists between those who either stake their position with evolution or some form of design or with creation. Some believe they can all co-exist.

When looking under the surface of the debate, it is clear much is at stake. Where we came from, our origins, says a great deal about who we are: how we educate ourselves, how we determine what is right and what is wrong, what we do with the poor and disenfranchised, even with how we determine which humans should live and which should die.

Consider the current demand by some evolutionists that Darwin's theory of evolution and only Darwin's theory of evolution be taught in the classroom. Are they, in effect, establishing themselves as the thought police in public schools with no challenges allowed? Many claim they are. Many are now calling for a level playing field. At the very minimum, they contend that the numerous unsupported hypotheses and contradictions of Darwinian evolution should be exposed in the classroom. Who is right?

Since the mid-1800s when Darwin postulated his theory, evolutionists have championed it. Paleontologist and evolutionist Niles Eldredge, Curator of Paleontology at the American Museum of Natural History, states the evolutionist view clearly. Though claiming some disparity with Darwin based on the "newer evolutionary science" of today, renowned evolutionist Eldredge says that all life on earth shares a common ancestor through the process of modification which, he claims, reveals that we are all distant cousins whether we are humans, plants, birds, or fish. As noted, in 2006, Niles Eldredge described himself as a "true Darwinist" in *The Virginia Quarterly Review*.[1]

Some evolutionists grant that if God does exist, they believe He does not intervene in the affairs of humans, nor care what we do or how we treat one another. Others like botanist Asa Gray, one of the earliest theistic evolutionists in the United States, was among the first to argue in the late 1850s that he believed evolution was compatible with religious belief.[2]

Recall that one of the key elements of evolution that has been proposed by followers of Darwin is that all humans evolved from ape-like ancestors. It is an idea that has long been printed in textbooks worldwide. The "ape-to-man" theory suggests that some humans may be more evolved or are more advanced along the evolutionary chain than others.

Is that a racist theory? Many believe it is. Former Assistant Secretary of State and U.S. Presidential candidate Alan Keyes is a student of both Social Darwinism and of evolution's impact on society. The following is a perspective from Keyes, a black American and Harvard graduate.

> *"...the prevalent defect of our governing elites, whose fancied sophistication has cut them off from the wisdom of America's Founding generation...think that He (GOD) is a convenient figment of human need and imagination, conjured up as humanity creates itself out of the chaos of material evolution. This may give comfort to human pride and arrogance, but it offers none to those who seek justice when the strong survive and dominate at the expense of human life, human dignity, and human freedom.*

Justice is not the good of the stronger. It is not the survival of the fittest. It is the universal birthright of all humanity, established not by our laws, not by our triumphs, and not even by our prayers, but by the will of the Creator.[3] **Alan Keyes**

At the heart of the debate is—is evolution <u>theory</u> or is it scientific <u>fact</u>? How do scientists reach their conclusions? Honest scientists have been and will always be of incalculable value to society. But, what about scientists who harbor pre-existing biases and reject all other source material that may lead them to the correct conclusions. Still there are others whose research is less than objective as they seek acceptance from colleagues or are driven by a need to justify a benefactor's grant money. The smartest scientists approach their work with a willingness to accept the facts wherever they lead. Whereas, others may postulate and hold rigidly to an outcome before the facts are in. Consider the following from science writer Boyce Rensberger in his book, *How the World Works*:

Most scientists first get their ideas about how the world works not through rigorously logical processes but through hunches and wild guesses. As individuals they often come to believe something to be true long before they assemble the hard evidence that will convince somebody else that it is. Motivated by faith in his own ideas and a desire for acceptance by his peers, a scientist will labor for years knowing in his heart that his theory is correct but devising experiment after experiment whose results he hopes will support his position.[4] **Boyce Rensberger**

If, as many scientists state that evolution is fact, should it not be incumbent upon them to provide evidence? Since evolutionist scientists have not been able to do so conclusively thus far, should they not open their minds to the findings of modern science that is today crumbling the foundations upon which evolution stands? The theory of evolution is tainted with racism, with the ideas that groups of people are less advanced and inferior to others.

Is it not time to openly discuss the possible ramifications of such teachings? Darwinism's use as a scientific justification for racism and genocide should be studied if for no other reason than to help insure that it doesn't happen again.

If Darwinism has provided support for racism, as many suggest, should Darwin's theory of evolution be taught in the classroom? American Civil Liberties Union Executive Director, Anthony Romero, holds that evolution is science. He and the ACLU opposed students learning about an Intelligent Design of the universe.

> *Our concern about this intelligent design approach, is that basically it's nothing more than religious teaching dressed up in a lab coat...they're teaching the students about where they came from, from a God.*[5] **Anthony Romero, ACLU**

The ACLU demands that the theory of evolution alone be taught to children at all levels of education. Some who oppose the ACLU, charge that teaching *evolution* is nothing more than teaching theory—a philosophy of naturalism dressed up in a lab coat. They counter that evolution is bankrupt as a scientific theory. That it is no more than an assumption based on a belief that is being fostered as fact.

Dr. Ian Taylor, a creationist scientist, born in the United Kingdom, disagrees that only evolution should be taught. Author of *In the Minds of Men: Darwin and the New World Order,* Taylor believes that Darwinism took hold of the educational system while no one was watching.

> *Ironically, there is no concrete evidence for evolution. I think we can say that quite emphatically. It is more in the minds of men than it is, actually, in fact out there in nature. Evolution permeates our society because it's taught in the public schools.*[6] **Ian Taylor**

Outspoken Civil Rights leader Reverend Al Sharpton is at odds with the ACLU when it comes to keeping God out of school.

> *As a minister I am for school prayer, but I am not for imposing prayer on schoolchildren. I believe that just as children are not forced to pray, they should not be forced from praying, if that's what they want to do. There should be a moment of silence to begin each school day, when children can either - say a silent prayer, meditate, or do nothing. That's not unconstitutional in my opinion. But to ban prayer from school is not only immoral, it is as wrong as forcing school prayer.*[7] **Reverend Al Sharpton**

American Civil Liberties Attorney Witold Walczak basically claims that God and evolution don't mix since "evolution is one of the pillars of modern biology."[8] He further stated as plaintiff's attorney in the 2005 Katzmiller vs. Dover Area School District trial in Harrisburg, Pennsylvania, that, those wanting to impose God, in the form of Intelligent Design, on school children are trying to "shackle our children's minds with fifteenth century pseudoscience."[9] Walczak also wants evolution to be taught as science.

The Dover case ruling by Judge Jones has been trumpeted far and wide as the death knell of Intelligent Design. It is important to question what caused all the fury since ID was not being taught in the classroom and no one had mandated it? All that students were told was that ID was another theory and there was a book in the library should they like to investigate it. It was up to the student to decide. Nothing was taught them or pushed on them.

Traipsing into Evolution by DeWolf, West, Luskin and Witt provides another view of the Dover decision. Their critique was fourfold: 1. Judge Jones wrongly assumed the authority to decide what science is. 2. Judge Jones conflated the question of whether something is scientific with the question of which scientific theory is *most popular*, 3. Judge Jones disqualified ID as science only by misrepresenting the facts, and 4. Judge Jones failed to treat religion in a neutral manner.[10]

As an outspoken opponent of creation being taught in the classroom, Eugenie Scott, Executive Director of the National Center for Science Education, calls it misleading and confusing to students to study both evolution and creationism. Many consider that an insult to the intelligence of our students.

So, what are we to make of such opposing views? One outcome of the ongoing debate is that it forces us to focus on what the real issue is. Fact: The study of evolution is the study of a theory. Theories should and must be challenged. It is the very nature of science to do so.

In other words, if Darwin's theory has yet to be proven as scientific fact, shouldn't science, itself, demand that Darwin's theory be challenged by sound scientific study? Which demands the next question: If Darwin's theory of evolution can survive the bright light of intelligent review based on advances in modern science, what are evolutionists so afraid of? According to top scientists around the world, the answer to that question is simple. Modern science has revealed cracks in the theory of evolution so large as to eventually disassemble it. Top scientists whose findings disagree with Darwin and evolution simply want their views to be heard and reviewed as well. To limit a study to one opinion, is not a study at all and it is certainly not scientific.

Stephen Meyer, a Senior Fellow at the Discovery Institute, puts it this way regarding a proposal that was before the Kansas State Board of Education:

> *What* [the Board] *is considering is a proposal that would allow students, yes, to learn about Darwin's theory in its full glory, but also to learn about the current scientific criticisms of the theory as they*

exist within the scientific literature.[11] **Stephen Meyer**

Another core issue inherent in the debate is a movement to eliminate God as Creator and leave man to his own devices and moral codes. If those who believe in God as Creator are willing to consider all research in the classroom and in the lab, why do so many evolutionists object to allowing opposing thought to get a balanced treatment? All who are seeking truth should want to protect scientific findings—even if they *disprove* a theory.

Grand Illusions author, Dr. George Grant believes that there is both a religious and a political agenda at work and that groups like the American Civil Liberties Union (ACLU) are at the heart of it.

> *The ACLU has a stake in the argument over Darwin. Simply because the ACLU has a very distinctive religious and political agenda rooted in a revolutionary worldview that comes from Darwinism. The whole intent of the ACLU is to change our society and make it into something new, something that the old Christian foundations of America never allow.*[12] **George Grant**

A serious review of both sides of the issue is not just a scientific issue, it is an educational issue. Not being allowed to present criticisms of evolution, allows only half of a debate. The principles of science in their correct and purest form should not be tossed out of the classroom door just because someone fears the mention of God. One of the greatest and most revered scientific geniuses of all time, Albert Einstein, spoke of God to his students when his research revealed to him an intelligence beyond the capacity of the human mind. If Einstein was not afraid, why are evolutionists?

As a black American who ran for the presidency, Alan Keyes has studied the importance of freedom of ideas:

But now in our schools there is a different ideology.
It isn't taught in the civics courses (I don't know
what we do teach in the civics courses), *but we*
teach it in the science classes. It masquerades as
science though it is taught as indoctrination. Last
time I looked you can question science; there is no
scientific theory that you're not allowed to
question.[13] **Alan Keyes**

Books are not yet physically being burned, but intellectual curiosity and the peoples' right to challenge a theory from the mid-1800s are being banished at the expense of current and future generations. What is more scientifically honest than allowing students and scientists to pursue truth with all of the evidence? Not to allow a balanced treatment in the science labs and in the classrooms is intellectually dishonest.

At its best, science is based on evidence, not on theory or the popularity of an idea. Conclusions that are not based on scientific fact may be wishful thinking or whimsy. Scientists who pronounce foregone conclusions, not backed by evidence, appear quick to reject any criticism or further investigation into their work. Since Darwin pronounced his findings before having the evidence, innumerable scientists, both past and present, denounce Darwin's theory as lacking credibility.

Cambrian Explosion

As discussed in the previous chapter, Darwin was concerned about the fossil record. One of Darwin's greatest fears was that a period of geological time known as the Cambrian Era could disprove his theory of natural evolution. Even in the nineteenth century, Darwin had knowledge that there was a sudden and inexplicable appearance of fossils in the earth's strata at essentially the same time. In that relatively short period of time known as the Cambrian Explosion or era—nearly all animal phyla that make up the major groups of animals, came into existence quite suddenly. It is sometimes called, "biology's big bang."

With this sudden burst of various life forms appearing across the earth, there was little if any record of why the sudden arrival of life occurred or where it came from. Nor did the major event leave behind evidence that identified which group is most related to the other. The earth's fossil record just shows that the life forms appeared suddenly and intact. Darwin wrote of his great concern of such evidence against his theory.

Today, if Darwin were alive, he would learn what modern science has discovered—that there were more fossils with an amazing sophistication of development that existed during the Cambrian Era than exist today. The rapid appearance of major groups of already complex animals suddenly appearing for the first time on earth, as evidenced by worldwide deposits in the fossil record, is powerful, indisputable proof against Darwin's theory that all things evolve over long periods of time through natural selection.

Even the avid evolutionist, Stephen Jay Gould, wrote the following in his book, *Wonderful Life* that the Burgess Shale Cambrian fossils include:

> ...*a wide range of disparity of anatomical design never again equaled, and not matched today by all the creatures in the world's oceans.*[14] **Stephen J. Gould**

With geological evidence that life, in a wide variety of forms, suddenly exploded into existence on this planet during the Cambrian age, Dr. Jonathan Wells challenges the broad statement that Darwin's theory somehow explains it. He calls such a claim:

> *"The biggest science-stopper in modern history."*[15]

A sudden appearance of life contradicts evolution expectations very dramatically. One MIT scientist who worked on dating the Cambrian fossil layers put in a telling dig to his biologist co-workers. "We now know how fast, fast is," grins Bowring, "And what I would like to ask my biologist friends is, how fast can evolution get before you start feeling uncomfortable?"[16] (Samuel Bowring, *Time Magazine*, 1995, p. 70.) What would further disillusion Darwin about his theory is that geologists say that since the unparalleled explosion of life on earth during the Cambrian Era, no new phyla or body form has emerged. Clearly, the Cambrian explosion that exists as physical documented fact of an event evident throughout the earth's strata—deals the severest of blows to the evolutionary theory. In other words, Cambrian fact trumps evolution theory.

Another thing that bothered Darwin about his theory was how to explain the complexity of the eye. If evolution is to be believed, it would indicate that the eye, slowly over time, evolved to what we have today. It was what Darwin believed. Going from the simple to the more complex was the crux of Darwin's theory. But, that theory crumbles with a lowly arthropod known as the trilobite.

Surely Darwin would be in awe were he alive to learn about today's findings about the extensively complex eye of that small creature. The trilobite first appeared in the early Cambrian period and has left behind extensive evidence of itself throughout the earth's fossil record.

Paleontology shows that during the Cambrian Explosion, the lowly trilobite appears to have come into existence with a highly complex set of eyes already intact. Extinct today, the trilobite has left behind a fossil record of its existence second only to the dinosaurs. There no ignoring the trilobite and its "fully functioning and complex holochroal eye."[17]

Sir Ernst Chain, co-recipient of the Nobel Prize for isolating and purifying penicillin, found the lack of scientific scrutiny of Darwin's theory unacceptable.

> *To postulate that the development and survival of the fittest is entirely a consequence of chance mutations seems to me a hypothesis based on no evidence and irreconcilable with the facts. These classical evolutionary theories are a gross over-simplification of an immensely complex and intricate mass of facts, and it amazes me that they are swallowed so uncritically and readily, and for such a long time, by so many scientists without a murmur of protest.*[18] **Ernst Chain**

Was there another motive by Darwin and his followers to advance a mere theory that had no proven scientific fact to support it? Was a desire to be superior to other humans their driving force? Or, was the Living God of the Bible their target?

According to Judeo-Christian belief, God created the universe and all that is in it. In doing so—God created the foundations of science, physics, chemistry, mathematics, gravity and astronomy along with all existing forms of energy and matter. For many of the world's best scientific minds, it is God's physical and biological laws and mysteries of the origins of the universe that they are striving to unravel and understand.

In The Beginning—Nothing Exploded

Then, there are evolutionists whose belief is in naturalism. At the core of that belief system is that all things came from nothing. To believe that all things come from nothing requires a giant leap of faith since the origin of the universe still remains one of the greatest mysteries facing all scientists today.

No one, to date, has proven inconclusively the "all things came from nothing theory." Astronomer and physicist, Dr.

Hugh Ross, has researched "quasars"—some of the most distant and ancient objects in the universe. Through extensive testing and research, Dr. Ross became convinced that the Bible is truly the Word of a Living God. According to Dr. Ross' biography, not all of his discoveries had to do with astrophysics. He observed with amazement the impact of describing for people both inside and outside his collegial community the process by which he came to personal faith in Jesus Christ. A few expressed dismay, but most seemed overjoyed to meet someone who started at (or near) religious ground zero and through reality testing, both scientific and historical, became convinced that the Bible is truly the Word of God. For his part, Ross was stunned to discover how many individuals believed— or disbelieved—blindly, without checking evidence.[19]

Even the man considered the preeminent physicist alive today, Dr. Stephen Hawking, is said to be "looking for a better model"[20] as he looks for the theory of everything.

Some Creation scientists, however, say their evolutionist colleagues need only to consider—that the universal Big Bang itself, had it occurred, could be the result of Intelligent Design or as creationists call it—an act of God. Dr. Ross considers himself a progressive Creationist and believes the earth is billions of years old. He has written the book titled, *Why the Universe Is the Way It Is*.

As a scientist, Ross openly states his belief in God as Creator and is convinced the universe was no accident. He contends that the Big Bang was engineered by God and that the physical laws of nature were put into place by God. What creationists say is certain is that creation had one witness at the beginning and that was its Creator— God. Humankind's attempts to fully comprehend, much less replicate, God's handiwork on earth and throughout the universe have thus far proven insurmountable.

As the Bible reveals, the best of our thoughts cannot compare to the least of God's.

BIBLE:

...My thoughts are not your thoughts, nor are your ways My ways, declares the Lord. For as the heavens are higher than the earth, so are My ways higher than your ways, and My thoughts than your thoughts. Isaiah 55:8–9[21]

Evolution scientists support the findings of one of their own, Herbert Spencer who found, that time, force, action, space and matter comprise all elements that exist in the universe. But, they needed only to look at the first verse of biblical scripture to see that the *Bible* identified these categories long before they did.

Revealed in the first verse of the *Bible* (Genesis 1:1) is evidence that God created all five categories before the existence of humankind.

*In the **beginning** (time) **God** (force) **created** (action) the **heavens** (space) and the **earth** (matter), which includes the **universe**.* (modified with emphasis added.[22])

Albert Einstein [23]

Albert Einstein, one of the superior minds in science, spoke at a conference at the Union Theological Seminary in New York and had the following to say on the subject:

Science can be created only by those who are thoroughly imbued with the aspiration toward truth and understanding. This source of feeling, however, springs from the sphere of religion...Science without religion is lame. Religion without science is blind.[24]
Albert Einstein, *Ideas And Opinions*, 1954.

Though a self-proclaimed agnostic, in 1927, Einstein reportedly wrote the following regarding a higher being:

My religiosity consists in a humble admiration of the illimitable superior spirit that reveals itself in the little that we, with our weak and transitory understanding, can comprehend of reality.[25] **Albert Einstein**

There was a period when some people had convinced themselves, and anyone who would listen, that the earth was flat. The *Bible,* on the other hand, clearly stated that the earth was <u>not</u> <u>flat</u>. In epochs of time before Magellan sailed around the globe, the *Bible* revealed to anyone who would read it—that the earth is **circular**, suspended in space and created by God. All of us need only to look at the book of Isaiah.

BIBLE
*"It is He (God) who sits above the **<u>circle</u>** of the earth..."* Isaiah 40:22[26] (emphasis added)

This biblical scripture, known as God's Word, was written almost 700 years before the birth of Christ. It clearly states that the earth is circular.

NASA PHOTO 27

The word "circle" means round—and "round" means circle. It could not be any clearer. The Bible described the earth as round while many still held the view that it was **flat**.

ROMANS 1:20 reveals that there is one reason the universe—with its galaxies, stars, planets, humans, animals, all other life, matter and energy—**exists**. This world and all that is in it was created by the unlimited power and intelligence of God. All people across the globe, in all corners of the world, have been made aware of God's creations.

BIBLE

"For since the creation of the world—God's invisible qualities—His eternal power and divine nature— have been clearly seen—being understood from what has been <u>made,</u> so that men are without excuse..." Romans 1:20[28] (emphasis added)

In an article in Creation Science Defense, Tom Carpenter wrote the following:

> *...it must be known also that many scientists and "smartest men" today accept and confirm the truth accuracy of the Bible. Warner von Braun developed many of the rockets used in the space program including the satellite "Explorer I" launched only four months after Sputnik. He was also heavily involved in the planning of all three manned space programs—Mercury, Gemini and Apollo.*[29]

Dr. Wernher von Braun, former Director of NASA, is regarded as the "Father of the American Space Program." He had this to say about the Creator of the universe:

> *There simply cannot be a creation without some kind of Spiritual Creator...in the world around us we can behold the obvious manifestation of the Divine plan of the Creator.*[30]

The Father of the American Space Program further wrote:

> *...the vast mysteries of the universe should only confirm our belief in the certainty of its Creator.*[31]

Nobel Laureate and American astrophysicist, Arno Penzias, shared the 1978 Nobel Prize with Robert Wilson for discovering cosmic background microwave radiation in space. Penzias' research into cosmology, the study of the universe, caused him to see evidence of a plan of divine creation. The astrophysicist had this to say about the accuracy of the Bible:

> *...the best* [scientific] *data we have are exactly what I would have predicted, had I had nothing to go on but the five books of Moses, the Psalms, the Bible as a whole.*[32]

Professor of Aerospace, Mechanical and Nuclear Engineering, at the University of Oklahoma, Dr. Edward Blick compares evolution to the flat-earth theory.

> *Evolution is a scientific fairy-tale just as the 'flat-earth theory' was in the 12th century.* [33]

Evolutionists' rejection of a scientific review of Darwin's theory of evolution is nothing new. Even in Darwin's day there were bold assertions that Darwin's findings were "fact" and were not to be challenged. That did not set well with one of the greatest preachers of that day, Charles Spurgeon. Upon entering the early debates in England, Spurgeon voiced then what many voice today. Spurgeon quickly noted a double standard that was pro-evolution and anti-creation and used sarcasm to criticize Darwin's unproven theory. In his monthly magazine, *The Sword and the Trowel*, Spurgeon wrote the following to all those who recognized God as Creator.

> **You** *are not to be dogmatic in theology,* <u>*but*</u> *for* **scientific men** *it is the correct thing.* **You** *are* <u>*never*</u> *to assert anything very strongly; but* <u>*scientists*</u> *may boldly assert* **what they cannot prove,** *which may demand a faith more credulous than* **any** *we possess.*[34] **Charles Spurgeon** (emphasis added)

Calling the theory of evolution unfounded, Spurgeon warned of the pitfalls of trusting pre-existing biases such as Darwin's that can both masquerade as and corrupt true science:

> *...you and I are to take our Bibles and shape and mould our belief according to the ever-shifting teachings of so-called scientific men. What folly is this! Why, the march of science, falsely so-called through the world may be traced by exploded fallacies and abandoned theories. Former explores once adored are now ridiculed; the continual wreckings of false hypothesis is a matter of universal notoriety. You may tell where the learned have encamped by the debris left behind of suppositions and theories as plentiful as broken bottles. As the quacks, who ruled the world of medicine in one age, are the scorn of the next, so has it been, and so will it be, with your atheistical savants and pretenders to science.*[35] **Charles Spurgeon**

Charles Spurgeon posed the question that so many ask today: Why are we told that Darwin's theory of evolution is a scientific <u>fact</u>—when Darwin himself called it a <u>theory</u>. Today, those who oppose it, call Darwin's theory mostly speculation, misrepresentation or falsification of the facts.

Again, enter Molecular and Cell Biologist Jonathan Wells, author of *Icons of Evolution.* Wells is among scientists who say that Darwin's theory of evolution conflicts with the evidence and should not be taught as fact in science classes. Wells says that Darwinian descent with modification by unguided natural processes—as a comprehensive explanation for all living things—is scientifically controversial until this day, because it does not fit the evidence.[36]

> *I do not embrace Darwinism, for the simple reason that Darwinism is false...By "Darwinism" I mean Charles Darwin's theory, in both its original and modern forms, that all living things are descended from a common ancestor and modified by unguided natural processes such as random mutation and survival of the fittest."* [37] **Jonathan Wells**

What does it all mean to people in their everyday lives? How does it affect them on a personal level whether they believe in Darwin or God? If Darwin is right and all living things came from a singular accident of nature, making all plants and animals our close relatives, what then does his theory of evolution say about "race"? If God created humans separate from the plants and animals, what then does God say about "race"? How does Darwinism differ from God? Modern science sheds light on these questions.

First, we should consider what we all have in common. The cell. The cell is not what Darwin thought it was. It is not an unspectacular, tiny blob of matter—a simple viscous glob from which all creatures advanced or "evolved" to a higher state of being. High-powered microscopes and advances in molecular biology that did not exist in Darwin's time, now tell us that Darwin was wrong. Science today reveals the tiny cell is composed of trillions of extraordinarily efficient molecular machines that function in a universe of their own within the human body. Tiny machines of such design that they elude the human capacity to replicate [38] in the laboratory.

In 1953, British molecular biologist Francis Crick, a prominent evolutionist, discovered the double helix structure of DNA, along with American James Watson. Deoxyribonucleic acid is a molecule that contains the genetic instructions used in the development and functioning of all known living organisms. Crick became known as the father of DNA. In a 2003 article, on Telegraph.co.uk, it was said that Crick's distaste for religion was "one of his prime motives in his work."[39]

But, before his death Crick wrote the following on page 88 of his book, *Life Itself, Its Origin and Nature*:

> *Every time I write a paper on the origins of life, I determine I will never write another one, because there is too much speculation running after too few facts.*[40] **Francis Crick**

Apparently frustrated by the lack of proof to support Darwin's theory of evolution, Dr. Crick further concluded:

> *An honest man, armed with all the knowledge available to us now, could only state that, in some sense, the origin of life appears at the moment to be almost a miracle.*[41] **Francis Crick**

A "miracle" came to the mind of the father of DNA. Secular scientist Francis Crick died in 2004 knowing that science had not yet discovered, nor could not duplicate, the origin of life. Nor has science been able to do so up until today. But, according to other scientists—Francis Crick, along with James Watson, did aid science in ways they may not have expected:

> *DNA has crushed the hopes of the biological evolutionists, for it provides clear evidence that every species is locked into its own coding pattern. It would be impossible for one species to change into another. It is a combination lock, and it is shut tight.*[42] **Aleksandr I. Oparin,** Soviet Biochemist, Founder of

the Biochemistry Institute by the USSR Academy of
Sciences

If DNA proves that the human code locks us into our own
species, then we are what we always were and will ever
be—human.

Creation scientists recognize that an intelligent power,
beyond human capability, is the source of this "miracle"
which Dr. Crick spoke of with such awe. The *Bible* says
this "miracle" is shared by all. Because we all came from
God's original human creations—we are all one blood.
God's words tell us that there is no basis, no rationale, for
racism. And, science confirms it.

On the other hand, Darwinism and evolution believe that
humanity got its start in a muddy pond that was struck
by lightning...a primordial soup, as some call it. With the
advent of Darwinism came a racial divide based on the
speculation that humans are not equal due to their
varying stages of development—that some remain
biologically closer to their ancestor, the ape.

The premise is that some people are more developed or
evolved, while others—judged by some evolutionists to be
more primal in appearance—are not yet as evolved. So, the
question again is: Does where we came from or the color of
our skin determine how we are treated as members of the
human race?

PART II

TODAY

H ow do we perceive one another today? Are we today breeding the "Darwin's Racists" of tomorrow? Are there indicators that we as a society need to be aware of if we are to protect the rights and freedoms of all?

CHAPTER 8

AMERICA: A BEACON OF FREEDOM

How individuals think about equality among the races has daily implications. But, what of those who serve in the most powerful positions on earth—are they evolutionists or creationists? As an example, American Presidents rule people of all so-called "races" and backgrounds. They take an oath to protect all citizens based on America's Constitution and its national Bill of Rights. The Constitution and the Bill of Rights represent powerful protections for all Americans. Built on biblical principles, they sustain us as a free nation.

The question is: Are our leaders evolutionists like Darwin, are they creationists who believe in God as Creator, or do they fall somewhere in between? Most American Presidents professed a belief in God.

Does the fuss about who we are or how we began really matter? If it mattered to Marx, Hitler and Sanger, should it matter to us? History records that all of them used an unproven "theory" as justification to oppress millions. Today, many want evolution to be taught as if it is fact, forcing any and all other study to be banned or excluded. Amid the gathering global criticism from scientists that evolution is highly flawed—doesn't true science demand that such critical analysis be taught alongside of Darwin's yet unproven theory?

Even Darwin demanded that if his "theory" of evolution could not be proven, then all of his speculations proved nothing and his theory would "break down."[1]

Many evolutionists reportedly want to keep creation or Intelligent Design out of the equation in the classroom and society. Should only one view be considered and others muffled or suppressed? Does freedom of thought matter? If evolutionists have gotten it wrong, shouldn't we know about it? Consider what Dr. Stephen Jay Gould, a professor of Geology and Paleontology at Harvard (deceased May, 2002), stated at a lecture regarding evolution:

> *For all practical purposes we're not evolving. There's no reason to think we're going to get bigger brains or smaller toes or whatever—we are what we are."[2]*
> **Stephen Jay Gould**

Dr. Gould shocked many of his pro-Darwin and pro-evolution colleagues when answering questions after a lecture when he reportedly stated:

> *The fossil record does not show gradual evolutionary changes and every paleontologist has known that since Cuvier.[3]* **Stephen Jay Gould**

Gould also had this to say in one of his essays about the fossil record and evolution:

> *Paleontologists have paid an exorbitant price for Darwin's argument. We fancy ourselves as the only true students of life's history, yet to preserve our favored account of evolution by natural selection we view our data as so bad that we never see the very process we profess to study.[4]*

If a well-known evolutionist had that to say about evolution, "we view our data as so bad that we never see the very process we profess to study"—that alone should open the door for critical thought and review of the deficiencies in Darwin's theory. Shouldn't freedom to study and research evolution without restrictions be mandated, particularly if evolution does not measure up among some of its strongest advocates?

As mentioned in an earlier chapter, over seven hundred scientists from across the globe have openly signed a statement questioning Darwinian evolution's ability to account for the complexity of life. Others, however, have refrained from signing for fear of repercussions to their careers. You may ask, isn't there academic freedom in this country? Shouldn't science progress by questioning even cherished theories and follow the evidence wherever it leads? The answer is, in theory, yes. But, in practicality, scientists are human too and often do not like it when favorite ideas are re-examined. Those who dare to step outside the bounds of dictated study and research often suffer the consequences.

Two other cases, including one that required a government subcommittee investigation, will be reviewed in later chapters. First, however, consider the case of an acclaimed scientist with an impressive record. Though well noted in the press, the case of Guillermo Gonzalez was explored in the 2008 documentary film, *Expelled,* with commentator Ben Stein.

Gonzalez, a planetary astronomer and associate professor at Iowa State University, has done research and taught at Iowa State for five years. Gonzalez has accumulated over sixty peer-reviewed publications in various science and astronomy journals. In addition, he has presented over twenty papers at scientific conferences and his work has been featured in such respected publications as *Science, Nature* and *Scientific American.*

Ordinarily, to become a tenured professor at research institutions, there are specific requirements that must be met. The Astronomy Department at Iowa State requires a minimum of fifteen research papers. Gonzalez should have felt quite secure since he published nearly five times that many papers. Gonzalez also co-authored an astronomy textbook through Cambridge University Press that he and others used at Iowa State. But, his initial application for tenure was denied. The faculty senate indicated his application was denied because he didn't meet certain necessary requirements.

However, many suspected he was denied tenure for his support for Intelligent Design through his popular book and film *The Privileged Planet*.[5] While having nothing to do with biological evolution, Gonzalez and his co-author Jay Richards maintain that our earth is not only uniquely suited for complex life, but is also amazingly well-suited for intelligent life to observe the cosmos. This dual purpose seems to suggest design.

In denying Gonzalez's initial appeal, the university president specifically stated the denial had nothing to do with Intelligent Design. Gonzalez further appealed to the University Board of Regents. In the meantime the Discovery Institute obtained internal university e-mails clearly indicating that the sole reason Gonzalez was denied tenure was due to his support of ID, despite the university's public denials. These e-mails reportedly also indicated that some of these university professors knew what they were doing was wrong and conspired to keep their deliberations secret.

The report revealed that the ISU Board of Regents refused to review this information or provide Gonzalez an opportunity to defend himself before they voted. Not surprisingly, Gonzalez's final appeal was denied in early February 2008. To view a full list of online and print articles and to view Gonzalez's academic record, visit the Discovery Institute's section on Gonzalez online at Discovery.org.[6]

This story is just the tip of the iceberg as these scenes are being played out not only in this country, but around the world. Exposure is an important step in seeing justice done. As mentioned, another case will be reviewed in chapter 15 that involves an investigation by the U.S. government of a public-funded government institution for discrimination and harassment.

Not all scientific institutions or labs demonstrate such discriminatory behavior. But, one has to ask why, in America particularly, are credentialed scientists harassed for challenging an orthodoxy? Such behavior is suspect in the highest degree. Parents and students alike should ask if the evolution lobby is hiding something. Clearly, there is more at stake than science.

Freedom is not a by-product of American society. It is a hard fought for Constitutional right. Are freedoms being infringed upon, including speech and religion? America was founded as a beacon of freedom for oppressed people. Today, it exists as the freest nation in recorded history. Rabbi Daniel Lapin, of Toward Tradition, serves on the board of the Jewish Policy Center in Washington, D.C. He sees America as a haven.

> *It is not an accident that America has provided the most tranquil level of prosperity that Jews have enjoyed for 2000 years and that's because America is a Christian nation—and not in spite of that. I think always of America's Bible belt as Judaism's safety belt... Only two governments in the world have been governed by a* (covenant) *constitution—ancient Israel and modern America.*[7] **Rabbi Daniel Lapin**

The world regards the Constitution of the United States to be the founding document of the freest country on the face of the earth. In it, its writers reflect their biblical ideas and beliefs. Foremost is the warning that America must always keep the "state"—meaning *government*—out of its places of worship. The Constitutional writers alerted all Americans to stand fast against and prevent state or government control of religion.

The term separation of church and state as premised by founding father Thomas Jefferson is intended to keep *government* from controlling religion but not, as some suggest, to keep religion out of government. Jefferson's words, in an 1802 letter to the Danbury Baptist Association, cannot be clearer. Jefferson warned against the government creating a *state religion* that would require all citizens to both attend in worship and support financially, such as the Church of England—from which our founders fled.

> *Believing with you that religion is a matter which lies solely between man and his God, that he owes account to none other for his faith or his worship, that the legislative powers of government reach actions only, and not opinions, I contemplate with sovereign reverence that act of the whole American people which declared that their legislature should "make no law respecting an establishment of religion, or prohibiting the free exercise thereof," thus building a wall of separation between church and State.*[8] **Thomas Jefferson**

Donald Lutz is a University of Houston history professor and Constitutional authority. Dr. Lutz notes that history records that the majority of our nation's founding fathers were devout Christians. Indeed, a great many of them were ministers.

Lutz contends that removing religion out of government was never their intention since it would have been in opposition to what they believed. Their intent was to practice religion without government interfering, as opposed to keeping religion out of the realm of government.[9]

According to Professor Lutz, the original and now current fifty state constitutions all acknowledge God. Dr. Lutz, author of *Colonial Origins of the American Constitution: A Documentary History*, says the Constitution is the product of a long American legal tradition that finds its origin in the *Bible*.

> *They came over with the* Bible. *It was the biblical idea of covenant making, brought by the Pilgrims and Puritans, which provided the intellectual foundation for the American "constitution-making*

tradition" that climaxed in the federal Constitution. The United States Constitution is the undeniable consequence of those biblical ideas.[10] **Donald Lutz**

Lincoln [11]

President Abraham Lincoln recognized the human failing to elevate self even above the God of the universe.

But we have forgotten God. We have forgotten the gracious Hand which preserved us in peace, and multiplied and enriched and strengthened us; and we have vainly imagined, in the deceitfulness of our hearts, that all these blessings were produced by some superior wisdom and virtue of our own.[12]
Abraham Lincoln

U.S. President Abraham Lincoln, a man much admired by other U.S. Presidents, clearly recognized the hand of God in the founding of America. For most Americans, Lincoln's comments remain relevant. They need only look at the headlines. Many say that an example of government misinterpreting what our forefathers intended is evidenced by a judge's ruling in Pennsylvania that tossed the study of Intelligent Design out of the Dover public school system. Federal Judge John E. Jones, III, ruled that Intelligent Design is not a science, but a religion, "Teaching ID to school children therefore violated the supposition to keep the church separate from state."[13]

The ruling was odd in that many consider that a *theory* being taught in schools—the theory of evolution—is itself a theory with its own quite religious believers and followers. They point to followers who seek to believe in and remain faithful to the theory even though science continues to question its validity. Those who consider the theory of evolution to be a false religion, object to being coerced to adhere to and worship that belief system against their will by government mandate.

It is that coercion against the will of the people that our founding fathers warned about in the First Amendment to the Constitution. In what is considered by many to be a mind-numbing ruling, they say the judge overrode the First Amendment through a misinterpretation of its original intent. The constitutional intent being: to prevent government from controlling religion. The founding fathers clearly did not intend to bar or eliminate faith in God from government.

Clearly, in their wisdom, America's founding fathers could perceive that someday a government under secular control would attempt to control religion, placing not only religious freedoms, but potentially all freedoms in jeopardy. It was not that long ago that Adolf Hitler conspired to take over and control the Judeo-Christian churches and synagogues in Germany and would have done so—had the world not stood up against his surge for power.

In America's *Declaration of Independence*, its writers proclaim that all people "are endowed by their Creator with certain inalienable rights." They may have even foreseen that organizations such as the ACLU would capitalize on the "freedoms of democracy" and use them in ways many Americans view as detrimental.

The American Civil Liberties Union was founded in 1920 by Roger Nash Baldwin. Baldwin made two trips to the [former] Soviet Union, and in 1928 published a book entitled *Liberty Under the Soviets*, which contained effusive praise for the communist USSR. In 1935, fifteen years after he co-founded the ACLU and with Stalin in full power and murdering his own people, Baldwin wrote:

> *I am for Socialism, disarmament and ultimately, for the abolishing of the State itself....I seek the social*

ownership of property, the abolition of the propertied class and sole control of those who produce wealth. Communism is the goal.[14] **Roger Nash Baldwin, Co-founder of the ACLU**

Along with Baldwin's, other voices were raised as they foresaw an apathetic America.

The American people will never knowingly adopt Socialism. But under the name of 'liberalism' they will adopt every fragment of the Socialist program, until one day America will be a Socialist nation, without knowing how it happened.[15] **Norman Thomas**, Socialist Party Presidential Candidate in 1940, 1944 and 1948, co-founder of the American Civil Liberties Union (ACLU).

We can't expect the American people to jump from capitalism to communism, but we can assist their elected leaders in giving them small doses of socialism, until they awaken one day to find that they have communism.[16] **Nikita Kruschev**, former Soviet Premier/dictator.

Does it matter what is being taught in classrooms, if contrasting ideas will be tolerated, or how information is presented by various news media? ACLU co-founder Norman Thomas and Soviet Dictator Nikita Kruschev thought so.

CHAPTER 9

FREEDOM OF THOUGHT vs. WHAT'S NOT BEING TAUGHT

"Science is the search for the truth."
Linus Pauling, Nobel Scientist

So wrote two-time Nobel Prize winning chemist Linus Pauling, a towering figure in twentieth century science. Pauling's comment is one of the most repeated and taught axioms in scientific circles. If science is the search for truth, then in the final analysis, it is evidence that reveals the truth. The big question about the theory of evolution that sits in the room like the 800-pound gorilla is: Where is the evidence? Is evolution science or wishful thinking?

> We are told dogmatically that Evolution is an established fact; but we are never told who has established it, and by what means. We are told, often enough, that the doctrine is founded upon evidence, and that indeed this evidence 'is henceforward above all verification...' but we are left entirely in the dark on the crucial question wherein, precisely, this evidence consists.[1] **Wolfgang Smith**

> ...many scientists and non-scientists have made Evolution into a religion, something to be defended against infidels...many students of biology—professors and textbook writers included—have been so carried away with the arguments for Evolution that they neglect to question it. They preach it...College students, having gone through such a closed system of education, themselves

become teachers, entering high schools to continue the process, using textbooks written by former classmates or professors. High standards of scholarship and teaching break down. Propaganda and the pursuit of power replace the pursuit of knowledge. Education becomes a fraud.[2] **George Kocan**, "Evolution Isn't Faith But Theory," *Chicago Tribune*, April 21, 1980. (abridged)

Should government controls be off-limits regarding not only the freedom of religion, as the First Amendment guarantees, but also the freedom to seek the truth in science and education? Is it ever right for truth to be censored?

Ron Paige, Former U.S. Secretary of Education, has seen the impact of taking God out of schools and public life. Schools today decry a reported drop in student safety and morals and a substantial increase in students who think it is okay to cheat and who are forced to accept violence as part of school life. Paige sees a lack of established standard values for the betterment of all.

In a religious environment the value system is set. That's not the case in a public school, where there are so many different kids with different kinds of values.[3]

Today, I ask students, teachers, parents and other proud Americans across the country to join me in showing our patriotism by reciting the Pledge of Allegiance at a single time and with a unified voice. Together, we can send a loud and powerful message that will be heard around the world: America is 'one nation, under God, indivisible, with liberty and justice for all.'[4]

If one is to believe headlines, the ACLU appears to be hard at work to reject God's moral influence, not only in schools and government, but also in society at large. The following are various headlines regarding the ACLU's activities:

- ACLU Intimidates and Seeks to Deny Free Speech to Christians
- ACLU Suit Behind Illinois House Dropping Support for Moment of Silence
- ACLU investigates Bible course at Virginia High School
- ACLU appeals federal decision in GA Prayer case
- ACLU urges Naval Academy to end lunchtime prayer.
- ACLU Defends Youthful Terrorist Bombers
- Judges Courtroom Poster of Ten Commandments Draws Fire from ACLU
- Indiana Judge Dismisses ACLU Challenge, Upholds 'God' License Plate
- ACLU asks Sacramento Library Board to relax pornography controls on public computers.
- ACLU pays to bus Nebraska College Students to South Dakota for Anti-life Rally and Canvassing.
- A blow to the ACLU—Supreme Court upholds a new federal law against the solicitation of child pornography.

And so on...[5]

Many Americans are surprised to learn that when the ACLU goes to court and challenges religious symbols, words or expressions and wins, those awards and ACLU attorneys' fees, have often been paid to the ACLU with taxpayers dollars. Targets have often been city governments, school systems and organizations like the Boy Scouts of America that are unable to afford to fight the ACLU in court.

The ACLU has systematically used such court cases to bring millions of the American public's dollars into its coffers. The irony being, some taxpayers who oppose the ACLU are in effect helping to fund its existence through the large awards being granted to the ACLU by American courts. Today, the ACLU and the courts appear to have a greater say and impact on morality and education than ever. Should the ACLU determine what can or cannot be mentioned in the nation's classrooms? Does the elimination of God supplanted by the teaching of Darwin impose a declining moral climate in our schools?

Clearly, a national confluence on morals can impact what is taught and what is not taught in schools. What is taught, and often what is not taught, can implant a lifelong mind-set for students. Which is why many ask— why is it that most Americans have gone through the entire educational system, including universities, without ever learning about the connection between Darwin's theory of evolution and racism?

Author of *Godless*, columnist and media commentator, Ann Coulter was stunned that it wasn't until she was an adult that she learned about the connection between Darwin and Adolf Hitler, one of the world's most renowned racists. Coulter found it incomprehensible that she could make it through twelve years of public school, then college and law school, and still not learn that it was Charles Darwin's theory of evolution that helped create the public sentiment that helped fuel Hitler's ovens.

> *I never knew about the link between Darwin and*
> *Hitler until after reading Richard Weikart's book.*[6]
> [From Darwin to Hitler, 2004]

Why is the Darwin and Hitler link so little known? Why has it been kept out of school curriculum? Nancy Pearcey—a Francis A. Schaeffer Scholar at the World Journalism Institute—is author of the book, *The Soul of Science*. Pearcey is an advocate for freedom of thought and believes that more education, not less, is needed. She wrote the following in her dust jacket review of Richard Weikart's 2004 book, *From Darwin to Hitler: Evolutionary Ethics, Eugenics, and Racism in Germany.*

> *If you think moral issues like infanticide, assisted*
> *suicide, and tampering with human genes are new,*
> *Weikart's book draws a chilling picture of how*

Darwinian naturalism led German thinkers to treat human life as raw materials to advance the course of evolution. Hitler's Germany was 'cutting edge' and in line with the scientific understanding of the day. Pearcey agrees with Weikart's implicit warning that as long as the same assumption of Darwinian naturalism reigns in educated circles in our own [current] day, it may well lead to similar practices.[7]

Pearcey's call for freedom of thought and a dismantling of the thought police also rang true with British evolutionist Professor H.S. Lipson after years of research. (*"A physicist looks at evolution"*...) In his Physics Bulletin at the University of Manchester in England, he came to the same conclusion that truth must out whether one likes it or not.

If living matter is not, then, caused by the interplay of atoms, natural forces and radiation, how has it come into being? I think, however, that we must go further than this and admit that the only acceptable explanation is creation. I know that this is anathema to physicists, as indeed it is to me, but we must not reject a theory that we do not like if the experimental evidence supports it.[8] **H.S. Lipson**

Much was publicized about the Columbine High School massacre outside of Denver in 1999. In a shooting rampage, students Eric Harris and Dylan Klebold reportedly murdered twelve students and a teacher and wounded twenty-three others. While the horrendous incident sparked a debate about gun control, no one much mentioned the issue of "mind control" in public schools. Few people heard about the Darwin connection to the shootings or what alleged shooter Eric Harris had reportedly written on his website:

You know what I love? NATURAL SELECTION. It's the best thing that ever happened to Earth. Getting rid of all the stupid and weak organisms.[9] **Eric Harris**

Harris and Klebold attacked on the anniversary of Adolf Hitler's birth date, April 20th. It was reportedly learned from Harris' autopsy report that he chose to wear a t-shirt with the words "Natural Selection" to signify his mental state during his murderous rampage.[10] Thirty-six peoples' lives ended or were changed that day as a result of such heinous thought.

With Columbine in mind, again consider the oft-quoted Richard Dawkins' *New York Times* article stating that: "It is absolutely safe to say that if you meet somebody who claims to not believe in evolution, that person is ignorant, stupid or insane, but I'd rather not print that)."[11]

Clearly, <u>evolutionary</u> thought can breed something far deadlier than alleged ignorance or stupidity. Columbine gives rise to a warning that bears repeating:

> *The philosophy that fueled German militarism and Hitlerism is taught as fact in every American public school, with no disagreement allowed. Every parent ought to know this story.*[12] **Phillip E. Johnson**

Test Of Time

It is always important to uncover what people like Darwin and his followers base their judgments of others on. How did they size up their fellow man and woman?

Living examples of why two of Darwin's advocates, Hitler and Sanger, got it wrong are Matt and Amy Roloff, stars of a television reality show *Little People, Big World*. Respectively, their professions have been entrepreneur and teacher. They are the happy parents of four children, three are average height and one is a little person like they are. Both sets of their own parents are average size.

The Roloffs are no different from everyone else whose physical make-up is determined by their genes. Some families have a history that affects height, weight, vision or a proclivity to one disease of another. It is all part of the human condition. The Roloff's genetic make-up includes dwarfism, yet all family offspring do not reflect the gene.

Matt Roloff is in demand around the country to speak on behalf of Little People and is an example of a true American success story. Both Amy and Matt agree that because of their physical differences, they have had to "adapt" to their environment. In fact, Matt has approached various hotel chains to include kits that will allow Little People to maneuver around their hotel accommodations with greater ease.

Matt & Amy drive re-designed cars and use special equipment to reach high places. They have sat atop Grand Canyon, enjoyed scuba diving and zip-lined over a deep crevasse for sport. Amy has played and coached soccer. Matt runs a thirty-four-acre farm and has designed and helped construct playground set pieces such as an old western town, a pirate ship and many other novelties, which have become major tourist attractions.[13]

Clearly, with the Roloffs, adaptation happens. Adapting to their environment makes their world more manageable. This, however, is not the kind of adaptation that evolutionists talk about. Evolutionists believe that plants, animals and humans "evolve" or mutate over great periods of time in response to changing environments. Evolution, after all, is going from a lesser state of being to an improved or greater state of being. Who determines who among their fellow human beings is lesser or greater?

Darwinians say society is pre-wired to be "dog-eat-dog" where only "the fittest" among us will survive. Hitler and Sanger would have categorized the Roloff's differences as a detriment to their ability to contribute to society. Having succeeded in life beyond what most people dream of, the Roloffs have clearly proven the Hitler, Sanger and Darwinian "survival of the fittest" mentality wrong!

George Foreman and George Foreman, III, are father and son. Would Darwin be amazed that George Foreman, an

American of African descent, had the determination and intellect to raise himself out of poverty? As a former heavy weight boxing champion of the world, George went on to become one of the most liked and successful business entrepreneurs of his day. His son, George III, graduated from the prestigious Pepperdine University and completed his master's at Rice University,[14] one of the most respected and academically challenging universities in the country. Today, their accomplishments are not an anomaly, but represent the vast numbers of all people who succeed. But, certainly as men of color, they too have proven Darwin, Hitler and Sanger wrong.

We all come in a variety of shapes and sizes as well as hair, eye and skin color—sometimes within the same families. It is often said from our own general observations of others, including our own families, that genes skip a generation or two. A child with a prominent nose and light hair may perplex others who notice that the parent's noses are an average size and that they both have dark hair. Until, they see a photo of the grandpapa and his sizable proboscis and light hair, which for most people—seems to explain it without need for scientific verification.

If parents are geniuses and the children are not, does that not fly in the face of evolution since we are all supposed to be evolving from a lesser state to a better state with only the best among us to survive? We are different, even in our own families, yet, we are all human. And, though you may think from time to time that your brother or first cousin is from another species, most assuredly they are not.

CHAPTER 10

MEDIA MATTERS

I n the newspaper business there is an old saying that "we print all the news that's fit to print". In recent times with lines blurring between tabloids, the Internet, and various media news organizations, who decides what is fit to print or report anymore? What oftentimes stands out more is what is not being reported. In other words, what is purposely being omitted and, thereby, pushed out of the public's consciousness?

Being in the news business is being in a position of power; power to sway public opinion, power to exclude opposing ideas, and power to flip old journalistic principles and standards on their heads. The public has watched as blatant opinion and agendas, being passed off as news, have weakened the fourth estate. The term *fourth estate* is used to describe the press in political context and position of power.

There is an affinity in most of the media for all things Darwin. Conscientious consumers of the media's daily output must ask: Is the whole truth being told so that the public can judge for itself? Or, is the public being propagandized with a pro-Darwin slant before the facts are in? How do we know? Is there a pattern?

Many ask, is the news media biased against God as Creator and pro-Darwin? Can evidence be found in the news media and broadcast television? Readers and viewers have observed an undeniable up-tic in pro-Darwin reporting and programming. Even a financial news network chimed in with a crawl that stated: "The Dow Theory is like Intelligent Design, it means nothing."

Journalists are people too. They are just as susceptible to prejudice and setting an agenda as anyone else. Many in the media seem to think their job is to tell us what to think, not just report what happened. What part does the media and television play in the creation vs. evolution debate?

The family centered organization, Focus on the Family, criticized ABC for choosing Holy Week, leading up to Easter, to "attack the legitimacy of Scriptures" and to try to "reduce the Word of God to myths." Focus says, the ABC special "The Search For Jesus", hosted by the late ABC anchorman Peter Jennings in 2000, was not successful in hiding its biases.

> *"Jennings repeatedly refers to 'the Jesus movement' as just another political party."*[1]

Again, Jennings was merely hosting the report. But, do networks make Christianity a prime target? The American Family Association documents television, news programs, and specials that repeatedly discredit religion. AFA's Founder and Chairman is Donald E. Wildmon.

> *ABC-TV was unwilling to present a scriptural concept of Jesus, choosing to give a distorted view with the impression that the New Testament cannot be trusted,"* AFA said on its website.[2] **Donald E. Wildmon**

There have been protests against current day attacks, like an ESPN anchor's reported use of profanity in comments regarding Jesus at a public event while representing ESPN.[3]

Why are the modern "mockers" so hostile to people of faith? Bernard Goldberg worked for CBS News for twenty-eight years and has won six Emmy Awards during his career. As a network insider, he witnessed media bias first hand.

In late 2001, Bernard Goldberg criticized the news media in his best-selling book, *Bias: A CBS Insider Exposes How the Media Distort the News.* Goldberg said that his experience at CBS demonstrated that the media is biased against conservatives and Christians.[4]

Why do so many news reports and television programs attempt to discredit or even re-write the Bible? In doing so, do they attempt to discredit God as Creator in an effort to put forth a pro-Darwin and pro-evolution position? The media often broadcasts "new discoveries" that are unfounded and often denigrate the Christian faith. Is it mere "entertainment" or does a large segment of the secular media have an evolutionary agenda to replace the God of the Bible with Darwin and his *Origin of Species*?

Time and again, media and pseudo documentary programming present nonsense such as Jesus was married and died in France. Dr. Darrell Bock, who has written twenty books on Christianity and is author of the *New York Times* best seller, "*Breaking The Da Vinci Code,*" calls the media's reporting of the non-biblical accounts "conversation stoppers".

Even if honest mistakes are made, are media groups too quick to jump on the pro-Darwin bandwagon? In 1999, the scientific resource magazine, *National Geographic,* found itself embroiled in fraudulent claims out of China regarding the so-called Archaeoraptor, a faked Dinosaur bird. The fossil had been glued together by a Chinese farmer. Before verifying the facts, NationalGeographic.com blared: "New Birdlike Dinosaurs From China Are True Missing Link."[5]

Soon after, headlines around the globe proclaimed the truth about the farmer's fraud and his accomplices.

Dinosaur bones were put together with the bones of a newer species of bird and they tried to pass it off as a very important new evolutionary intermediate.[6]

Harsh response sprang from many quarters including the prestigious Smithsonian Institution's Curator of Birds, Storrs L. Olson. In regard to its article, Olson wrote an open letter to the National Geographic Society saying:

"National Geographic has reached an all-time low for engaging in sensationalistic, unsubstantiated, tabloid journalism...Sloan's article takes the prejudice to an entirely new level that makes news rather than reporting it...Truth and careful weighing of evidence has been among the first casualties in their program which is now fast becoming one of the grander scientific hoaxes of our age..." **Storrs L. Olson,** Curator of Birds, National Museum of Natural History, Smithsonian Museum, Washington, D.C.[7]

Why was the magazine so quick to print a story that turned out to be based on a fraud? Were they following the lead of Thomas Huxley, a friend of Charles Darwin, who first set forth the idea that birds and dinosaurs were connected? Or, is evolution their creed? One-sided reporting, or reporting with unsubstantiated information, is supposed to be an anathema, the first major sin against all journalistic ethics and principles. Even Darwin himself, the man behind the theory of evolution, had this to say about the importance of considering various sides of an issue?

A fair result can only be obtained by fully stating and balancing the facts and arguments on both sides of the question. **Charles Darwin**[8]

Crimes Against Humanity

Faked bird stories are one thing, but has the press also set its sights on people who believe in God as Creator of the universe? What the press often fails to report is that true religion, as practiced according to God's word, teaches truth and tolerance. With what he calls almost seemingly endless attacks on religion—through lawsuits, misleading television programs and a bias towards the faith of Darwinism, Author and President of Amerisearch, William J. Federer sees an alarming irony:

> *Religion provides social order and tolerance for all people. Now, in America, everyone is tolerated except the ones who came up with the idea.*"⁹ (emphasis added)

When the tolerant are not tolerated, what is the result? The pattern appeared before, under the Nazi Regime in Germany during World War II, when the German Supreme Court ruled segments of their population were "non-humans". Such actions by a governing body were the result of that nation's spiritual poverty and intolerance.

One need only recall Stalin's Russia and its elimination of God from society. As an atheist, Stalin's intent was to create a Godless society. In one of the bloodiest regimes in history, Stalin had millions of his own people executed. In that environment of dread and fear, Stalin promoted Darwin's theory over God. Stalin's intent was to separate people from God and force them to look to the state. Communism and its government leaders would then become their gods.

Landmarks in the Life of Stalin was a book written by E. Yaroslavsky to promote Stalin. The following pro-Stalin propaganda, published in Moscow in 1940, illustrates Joseph Stalin's anti-God and pro-Darwin mind-set:

> "*I'll lend you a book to read; it will show you that the world and all living things are quite different from what you imagine, and all this talk about God is sheer nonsense,*" *Joseph said.*
> "*What book is that?*" *I enquired.*
> "*Darwin. You must read it,*" *Joseph impressed on me.*"¹⁰

As noted, Joseph Stalin stands as history's worst mass murderer. But, after the advent of Darwinism, it was former Harvard Professor Steven Gould who observed that racism exploded with Darwinism. As did the slaughter and deaths of millions. History records that the bloodiest century in world history was the result of God-denying dictators such as Stalin, Hitler, Mao and Pol Pot.

History records that World War II was the single deadliest war the world has ever seen. Hitler's dream of a super-race thrust him into a worldwide conflict in his quest for superiority and power. Using Darwinian evolution as his blueprint, the German Fuhrer justified his and his allies' actions. An estimated 50 to 70 million military and civilian lives were lost in Hitler's failed attempt to conquer the world.

History also tells us that people have been killed in the name or pretense of religion. In regard to so-called religious wars, man has long usurped the name of God for his own benefit. History reveals that wars can be fought in the NAME of anything. From land grabs to personal agendas, using Christianity or any faith as a front to cover-up an individual's or a nation's quest for power or territory is nothing new.

Any cause, including race and religion, can be used by unscrupulous leaders as their "front" or "guise" to do injustice or gain power over others. God allows His people to protect themselves and defend their faith. But, acting outside the righteous authority of God—is man's doing, not God's. America has a history of white fringe groups claiming that they are Christian as they pursue racial superiority. Yet, the Bible instructs that racism is a sin as it commands us to love one another. Black rights organizations of the 1960s and 1970s originally sought to pursue black nationalism and racial superiority as they rejected the idea of non-violent integration as proposed by the Reverend Martin Luther King, Jr.

Religion and just causes may teach tolerance and non-violence, but it is man who falsely uses religion or a righteous cause as a means to hide his own bigotry, ambition or wrong doings. One need only consider the actions, in 2001, of twelve extremists who flew planes into New York's World Trade Center under the guise of Islam. Renowned author, Jonah Goldberg, author of the book, *Liberal Fascism,* notes that bigotry is a universal problem that afflicts all people—because it is perpetrated by all people.

When religion is not practiced according to God's word—humankind breaks down and fails itself. Wisdom is learning from the past. How we view ourselves today through the fractured lens of the media may tell us about our future.

PART III

TOMORROW

With the future moving quickly towards us, will we learn from the past, or will history repeat itself? As science and technology progress, society is experiencing rapid changes. At this point in our history it is important to ask ourselves, what will be the impact of God vs. Darwin on both race and class in the future?

CHAPTER 11

CLONING & GENETIC ENGINEERING

Modern science has already provided insight into the future with the advent of <u>CLONING</u>.

CLONING:
...the process of making a clone, a genetically identical copy.[1]

What is a clone, who is a clone, and why would someone want a clone? How will society be impacted by laboratory produced beings or as some call them "human inventions" that are designed to serve either society's purpose or an individual's need? Who will science clone? Will one group be cloned more than another? Will people of color be included? Who decides?

It is no longer shocking to the public to hear about someone spending tens of thousands of dollars to clone a beloved pet. Always at question, however, is how healthy will the clone be physically or mentally. Disabilities or inner deformities may not be immediately evident. Though a clone may appear similar, it is not a twin.

Cloning does not happen by accident or chance. Cloning is the purposeful attempt to make copies, as best science can, of existing life forms. It can only happen or be accomplished through intelligence as applied by the human mind on living species or tissue. Cloning does not and cannot create life. Science, even in its most advanced state, cannot *create* life. Cloning is the attempted duplication of pre-existing life.

The first adult mammal cloned was Dolly the sheep in 1997 at The Roslin Institute in the United Kingdom. Cloning often requires that a number of living embryos be used in order to increase the odds of success. That means for human cloning, it will be necessary to kill several living embryos to obtain the necessary DNA. Is it important to consider that were "the human embryos" allowed to develop naturally, they may grow up to be someone's sibling, neighbor or friend? With cloning, however, most will end up in a trash bin as "waste material" and forgotten?

The Bible tells us to value human life and that we are all God's creations.

> *For You formed my inward parts; You wove me in my mother's womb. I will give thanks to You, because I am fearfully and wonderfully made; Wonderful are Your works, and my soul knows it very well. My frame was not hidden from You, When I was made in secret, And skillfully wrought in the depths of the earth. Your eyes have seen my unformed substance; And in your book were all written The days that were ordained for me, When as yet there was not one of them.[2] Psalm 139:13–16*

The Bible informs that as your Creator, God knows you before you are born and at the very moment of conception you are a human being with a future. History already records that without God as the ultimate authority in our minds and hearts, mankind has proven its ability to devalue life, dehumanize his fellow man and woman, and discard life as if it were waste material.

Are we to believe, that the same racist mind-set, based on the evolutionary standards used by the Hitlers of the world, will somehow be avoided when it involves cloning? Left in the hands of godless men and women, one can expect the worst of history to repeat itself...in startling new ways.

Cloning will leave many disappointed. Much of the excitement about cloning is based primarily on unrealistic beliefs. Some people will clone themselves as a vanity-product. Other people think if they clone themselves they will live forever. Others believe they can produce an exact replica of a lost child or pet. Scientists, however, are quick to explain that if Adolf Hitler were cloned, you may end up with Billy Graham or vice-a-versa. In other words, what you set out to clone - is not what you get. With today's vast array of other options available, people don't need cloning to have a child. If a couple chooses to clone, motives should certainly be questioned.

So, who benefits from cloning? Certainly not the being that is cloned. Since the cloned person is not a twin to the tissue it was cloned from, scientists say that the clone's DNA is far from being identical as well. The clone's newly designed DNA is highly unpredictable with potentially erratic results. A clone may suffer greatly with serious complications. Will those who deem themselves superior even care about a cloned being? Will those who order up a clone with the swipe of a credit card regard them and treat them as humans with equal rights and freedoms?

Today, cloning to make a baby is illegal. Yet, researchers and laboratories on the fringes are reportedly attempting to do just that. One underground lab has claimed it has succeeded in cloning a baby, but there is no evidence to support that claim. In Great Britain, scientists sought governmental approval to use and destroy living embryos if there are no other means to meet an experiment's objectives. Scientists warn that the consequences of tinkering are devastating. The true downside of cloning is that it has resulted in debilitating deformities, diseases and deaths of animals in the lab. The outcome is said to be extremely unpredictable. Serious deformities and painful deaths occur with cloning such as Dolly the sheep aging quickly and having severe arthritis.

The risks to the cloned individual are enormous. But, again, any benefits to the cloned individual are not really part of the equation. Harsh reality is that the welfare of the clone is expected to have limited to no relevance in most cases. The benefits are to the people who want to own or have a clone.

Having stated his support of the tremendous progress of genetic medicine, Former U.S. President George W. Bush said, when signing a law outlawing cloning in the United States that, "Life is a creation, not a commodity".[3] Was he right? Setting all political views aside; does how we answer that—reflect <u>who</u> we are as a society today? Cloning is a divisive issue with foreboding societal implications in the future.

How will the human race handle cloning? Once perfected in the laboratory, will cloning be based on the Darwinian formula of the survival of the fittest—weeding out the weak, the poor, and those who do not meet its standards? Will the color of one's skin be a determining factor if they will be cloned? Clearly, with his visions of a "super-race", cloning would have been Adolf Hitler's dream come true. Hitler would have co-opted the Darwinian model to eliminate everyone except those he deemed superior and worthy to live.

Will cloning create an ever-widening separation of people? Will some people be cloned for the singular purpose to serve others as a slave class? What laws will be put into place to prevent society from repeating its sins of the past? If future generations deny God and are brought up and schooled only on evolution and its core racist' idea of superior beings—will it care or even know how to restrict itself?

As individuals and as a society we not only seek, but fight for cures to diseases that devastate humanity. Diseases that fell both the supposedly "fit" and the "unfit" alike. But, according to a U.S. Senate subcommittee hearing report on Science, Technology and Space, cures from therapeutic cloning will not be "generally available" to all of us as a medical treatment. At issue is therapeutic cloning's potential risk to women's health since it requires harvesting vast numbers of human eggs.

In his introduction to the hearing, Senator Sam Brownback of Kansas quoted the following estimation from a medical specialist: "To get enough eggs to seek clone cures for these four diseases—just four diseases— LLS, Parkinson's, Alzheimer's, and diabetes, every woman in the U.S. aged 18 to 44—that is approximately fifty-five million—would have to endure two cycles of ovarian hormone hyper-stimulation and then undergo surgery." Brownback then quoted an executive of Geron Corporation as saying it would "take thousands of human eggs on an assembly line" to produce a custom therapy for a single person. The exec further stated, "This is a non-starter, commercially." Others voiced concerns about the rights of women and the potential for wholesale marketing for eggs that would have millions of women undergoing health threatening hormone treatments and surgery.

> The entire transcript of the March 27, 2003, U.S. Senate subcommittee hearing can be read at: http://bulk.resource.org/gpo.gov/hearings/108s/96697.pdf

Were it to become viable in the future, at first glance cloning appears to be a benevolent cure for what ails the human race. But, history shows that all good intentions can go terribly wrong if they fall into the wrong hands. In the long run, will human cloning bring benefit to some and suffering to others? It is both foolish and reckless to only look at cloning as a mere medical advancement and ignore its potentially sexist and racist dark side. As in all things—absence good people, evil prevails.

Jumping The Code

As cloning becomes a part of the cultural vernacular, a lesser-known but possibly more revolutionary scientific advancement known as Genetic Engineering provides a glimpse into our future. In short, Genetic Engineering is the "re-engineering" of ourselves from the inside out by manipulating or altering our genes. Genetic Engineering is a prime example of a change or advancement that would never occur without an external intelligence taking specific actions—in a laboratory—to make those changes happen. The intelligence that drives genetic engineering is, of course, human intelligence. Manipulation of a cell's DNA is done with purpose and forethought.

Clearly, such premeditated change is in direct contrast to the evolutionary process of chance. Evolution is described as occurring without plan or purpose and occurring by random chance. Evolutionary biologists would have us believe that over time through natural processes organisms evolve or jump outside their species.

Clearly, in Darwin's day, he did not have access to the advanced equipment and technology of today. Darwin based his observations primarily on outer appearances and not an organism's cellular or molecular structure. He primarily drew his conclusions on what he could see through his magnifying glass or rudimentary microscope. Darwin was conflicted about his theory of evolution and wrote about his nagging doubts.

He may have suspected that one day, science might prove him wrong. Had Darwin known about DNA, he may have never proffered up his theory. What would Darwin learn from DNA were he alive today? First, he would learn that very specific encoded information exists in the DNA of every living organism.

Take for example the microscopic amoeba. The single cell amoeba is a microscopic mass of translucent, semi-fluid matter called protoplasm, a Jell-O-like substance filled with even smaller cellular machines made of proteins. Being able to examine the single cell amoeba was beyond Darwin's capability since he had neither advanced equipment nor the scientific knowledge to do so. All that was available to Darwin in the 1800s were very low powered microscopes.

What science has learned that Darwin could not know is that every living thing, including the tiny amoeba, has its own code or set of molecular instructions within its DNA. Darwin, and all scientists of his day, believed that cells were a simple form of life with no complexity. From that simplest form of life, he theorized that all life evolved to a higher or more complex state. Modern science, through molecular and cellular biology, has proven Darwin's simplistic assumptions wrong.

Darwin's unsubstantiated theory has no validity when overwhelmed by the molecular evidence of today. Darwin's is not the first theory that has failed as science has advanced. It was, after all, the year 1859 when he introduced his theory. Again, Darwin relied primarily on appearances and his limited ability to make sense of what he was seeing without the benefit of modern science.

Today, DNA tells us that all genetic matter, humans included, has its own very specific set of instructions. DNA would today inform Darwin that cells are not simple. Indeed, they are remarkably complex and highly efficient machines that both repair and reproduce themselves. Although DNA in all living things may have some similar characteristics, each species has its own uniquely specific set of DNA instructions. Today, modern genetics would inform Darwin that an amoeba cell, locked into and restricted by its very specific DNA code, can only produce other amoeba. Just like an oak tree's DNA instructions can only produce other oak trees. Genetics would inform Darwin, presumably much to his amazement, that a bee's genetic instructions only permit the bee to reproduce other bees.

Microevolution is without controversy and was known about and recognized long before Darwin came along. It has long been accepted that plants, animals and humans may experience minor changes or "adaptations" in response to variances in the environment—such as lighter skin in colder climates, which enhances Vitamin D production—as opposed to warmer climates where darker skin protects against excess sun damage. But, these types of adaptations do not <u>change</u> one species into another. In scientific terms, small changes—<u>micro</u>evolution—do not cause major changes as in turning a monkey or some purported ape-like ancestor into a human.

That magnitude of change, which has never occurred, would be called <u>macro</u>evolution. In other words, Darwin's finches, as mentioned in an earlier chapter, are all still finches. Their beak sizes and shapes may change slightly to accommodate different sizes and shapes of seeds from plants during either extreme periods of drought or rain. But, through it all, the finches still remain finches.

The Monkey's Dilemma

Modern molecular and cell biology support that life was formed and exists, not by chance, but by design. The world's varied species and diverse groups are distinct from one another, and though some adaptive variations can occur within any given group itself, there is no conclusive evidence that macroevolution has ever happened. In other words, there is no convincing evidence that a monkey was ever your uncle, or anything else but a monkey—or that your life began in a mud puddle.

Charles Darwin depicted as an ape, Hornet Magazine, 1871.[4]

But, since there has been much to do about the supposed ape-like ancestry of humans, it deserves a closer look. Darwin believed that some groups of darker skinned people were closer to "their ape ancestry" than others and are, as a consequence, less evolved. Such ideas follow his theory of natural selection. But, DNA would today tell Darwin that chimpanzees, or humankinds' presumed ape-like ancestors, cannot jump the code. The chimpanzee's DNA code makes the chimp behave like a chimp—and reproduce chimps—but it never turns a chimpanzee or any other supposed ape-like ancestor into a man. DNA shows that there is no such thing as jumping out of your species—or jumping the code. If monkeys, chimps or apes evolved into man, the transitional groups would be witnessed among us today in their various stages. Indeed, there is no verifiable, documented evidence of the existence of such transitional creatures in the history of the world.

Bio-chemist Dr. Fazale Rana, Vice President of Reasons to Believe in Southern California, has made the study of DNA part of his life's work:

> *Scientists still lack a clear understanding of the genetic similarities and differences between humans and chimpanzees. But as the comparisons move from single genes to larger regions of the genome, researchers are exposing substantial distinctions. Humans and chimpanzees just don't prove as genetically similar as some once thought...*[5]

Monkeys make monkeys and amoebas make amoebas. If the single cell amoeba could make anything other than single cell amoebas, it would be witnessed in laboratories around the world. Though there is a lot of activity going on inside, an amoeba does not have the genetic or DNA capacity to independently transform itself from being a single cell. It cannot alone self-design kidneys, a liver, the heart, or a complex set of eyes.

The bottom line is: "There are cells that are organisms unto themselves, such as microscopic amoeba and bacteria." However, as Creationist and Theologian, Dr. John MacArthur says, "Some have chosen to believe in the eternal amoeba over the eternal God." No moral or ethical laws exist within the amoeba. Nor does the foundation for the existence of a conscience, emotion or insight evolve from the tiny viscous glob. According to modern science, the amoeba's DNA code does not allow it. Since DNA disproves the existence of an "amoeba creator" based on the evolutionary concepts of chance and natural selection, then what is the answer? MacArthur says that an Intelligent Designer not only created humankind, but also instilled the inherent knowledge of right and wrong that makes all human beings accountable to the Creator God. He adds, that the more science learns about design through DNA and other scientific study; it is confronted with that Intelligent Designer—God the Creator.[7]

In his book, "*Darwin Was Wrong: A Study in Probabilities*," Dr. I.L. Cohen put it in scientific terms that Darwin could understand had he been armed with the evidence of DNA.

> '*Survival of the fittest*' and '*natural selection.*' No matter what phraseology one generates, the basic fact remains the same: any physical change of any size, shape or form is strictly the result of purposeful alignment of billions of nucleotides (in the DNA). Nature or species do not have the capacity for rearranging them, nor adding to them. Consequently no leap (saltation) can occur from one species to another. The only way we know for DNA to be altered is through a meaningful intervention from an outside source of intelligence: one who knows what it is doing, such as our genetic engineers are now performing in their laboratories.[8] **I.L. Cohen**

In other words, it does not happen by chance. There is, however, another player in this debate. The lowly fruit fly.

> It is a striking, but not much mentioned fact that, though geneticists have been breeding fruit-flies for sixty years or more in labs all round the world—flies which produce a new generation every eleven days—they have never yet seen the emergence of a new species or even a new enzyme.[9] **Gordon Rattray Taylor, former Chief Science Advisor, BBC Television.**

As the small insect repeatedly produces a new generation every eleven days, no one has been able to command the fruit fly—"Fruit fly, you've had incalculable *chances* to change, so MUTATE already!" After unknown years of existence on this earth, the fruit fly has neither the capacity to understand the command nor the ability to evolve or mutate into something else. The fruit fly has repeatedly proven that it is locked into its own genetic code. It is locked into its own DNA that instructs what it is, has been and forever will be.

According to a Nobel Prize Laureate of Medicine, the fruit fly has been better off not being able to mutate into something else. Just seventy-eight years after Darwin introduced his theory, it was Dr. Albert Szent-Gyorgi who stated the following:

> *To improve a living organism by random mutation is like saying you could improve a Swiss watch by dropping it and bending one of its wheels or axis. Improving life by random mutations has the probability of zero.*[10]

As Dr. Jonathan Wells so succinctly put it in this oft-repeated quote:

> *All the evidence points to one conclusion: no matter what we do to a fruit fly embryo, there are only three possible outcomes—a normal fruit fly, a defective fruit fly or a dead fruit fly—not even a horse fly, much less a horse.*[11]

There is no dispute that changes or adaptations to plants, animals and even humans may occur over great periods of time. That happens when a species has to adapt to a changed environment or some major external pressure. To repeat, when changes occur to a plant or animal within its own species it is called micro-evolution. Micro refers to small changes. There is no big leap from one long-existing species into a whole other or new species. Were that to ever happen, it would be called macro-evolution.

Macro-evolution is the premise for Darwin's whole theory. Darwin named his book *On The Origin of Species* in which he expresses his belief that a new species can completely originate out of another—such as a human evolving or jumping the code from an ape.

As Dr. Jonathan Wells points out:

> *A species can change within itself, making minor changes. It cannot change to an entirely different animal or different species. Micro-evolution can be found in abundance.*[12]

Scientists will tell you that the jump from microevolution to macroevolution has never happened in any species. According to Dr. Ian Taylor, macroevolution or species jumping, has never been found.

> *What Darwin proposed was that any creature given enough time and different circumstances would grow into another creature. That's macro-evolution. Never been any evidence for it. Micro-evolution is just the variation within the species.*[13]

The odds of accidentally producing the correct DNA code in a species or changing it into another are mathematically impossible. [14]

As brilliant minds seek answers, they may reach divergent conclusions. Ongoing research in laboratories, in the effort to either support or deny God, has led scientists to amazing discoveries that benefit the human race. However, common to all scientific pursuits is that no one, in any lab at any time, has ever discovered how to "make life" out of nothing.

Genetic Engineering is not the making or creation of life, but the reworking or manipulation of already existing forms of life. In The Journal of the American Medical Association (JAMA), Francis Collins, former head of the futuristic sounding—Human Genome Project, and Alan E. Guttmacher wrote:

...integration of genomics into medical research and medical practice will be revolutionary...By providing more sophisticated knowledge of biology at the individual level and of disease typology, genomics has begun to personalize health care. By identifying those people for whom a specific drug is likely to be of benefit or harm, genomics has already begun to expand the pharmacopoeia. By providing new medical applications to the developing world, genomics has already begun to affect global health.[15]

Super Babies

In early 2000 came the startling news that scientists had injected mice embryos with the gene that makes a jellyfish become fluorescent. When the mice were born and put under a fluorescent light they reportedly emitted a green glow. Crossing a jellyfish with a mouse may sound like fun science to some, but many of the mice were born deformed and defective. Can the same outcome happen to humans?

What does today's genetic research data tell us about our future? As with any new scientific frontier, genetic engineering brings great hope and great challenges. It is always human nature to push the envelope. On the plus side, Genetic Engineering is expected to allow many to be spared lives of debilitating pain and illnesses.

Transhumanists are among those who welcome with open arms the possibilities of genetic engineering to alter who and what we are. They scoff at the reluctance of others to step into this coming Brave New World. They relish the possibilities of double and triple average life-expectancy, designer babies, and the elimination of genetic disease.

They aren't troubled by the necessity of costly mistakes and failures. That's just the price of research and progress. We accept risk all the time they say, why should genetic research be any different. They apply rather consistently a naturalistic worldview which sees human beings as just another species.

New frontiers, however, are filled with the unknown. Some speculate that genetic engineering will not only be used to improve the health of the human race, but that an elite, wealthy class may create a schism in society that will separate the haves from the have-nots. It is expected that some will be able to afford to buy everything from genetically engineered designer children to increased athletic ability, higher IQs, memory enhancement, better physiques and a more beautiful appearance—in other words taller, stronger, smarter, and better.

Engineered fetuses or 'designer babies', as they have been dubbed, will not be within most people's reach. Even so, talk of super babies and the technology to produce them periodically pops up in the headlines and has permeated conferences for the American Society for Reproductive Medicine. The future price tag will be costly for anyone willing to pay for a customized, "superior" baby. Major profits are expected for labs willing to alter and sell genetically enhanced fetuses.

Designer children, bionic athletes, rare beauty, and genetically designed Einsteins sound intriguing, but their genetic manufacturing raises ethical questions. When only part of the population can afford such superior attributes, will those who cannot be relegated to second-class status? What will genetic engineering's impact be on race, class and disability?

Only a few years into the genome era, the promise of genomics has already begun to be realized. Within its promise exists the potential for both good and bad. A process already in use called pre-implantation genetic diagnosis or PGD, not only screens for genetic diseases, but also allows parents to select the gender of their baby. Such embryo screening has been in use for almost twenty years. Some labs claim to have an increased probability that they can design a baby of choice. The future potential to pick eye, hair and skin color may seem innocuous to some, but others find it disturbing. There are concerns that parents seeking to mold and control their children's physical and mental existence through genetics pose serious societal implications.

Trying to alter ourselves genetically is probably inevitable because the technology is developing rapidly using animal models. And whatever we have done in animals, we eventually do in humans. The Darwinian worldview says quite strongly that we are just another animal species.

What of normal birth children whose genes don't measure up to the new ideal? Will they be subjected to potential discrimination because their genes may foretell of debilitating health issues? Serious ethical questions are raised and must be dealt with lest the future becomes a genetic wild west with no controls.

Cloning and genetic engineering represent a significant change in health care along with the future experimentation with carte blanche babies. The medical profession recognizes they can have a positive or negative impact depending on the public's response to such life altering issues. Clearly, laws must be established to direct use so that all of society may be served. But whose worldview will be used to develop these laws? If the Darwinian worldview holds sway, then we are all at risk for being deemed not good enough reproductive material to enhance the survival of the species.

So, again, who will decide who is fit and who is unfit, who will be the final arbiters of life? At present, life and death decisions are being made by governments worldwide; particularly those in countries with socialized medicine. Tragic reports have surfaced for years regarding patients who could not get government approval soon enough for a treatment that would have been life saving. The race between bureaucratic delays and impending death have reportedly resulted in men and women dying in markedly higher percentages from various cancers in countries with socialized medicine versus a considerably higher survival rate in the United States. Patients reportedly have gone through great anguish waiting on government approval to get treatment before they finally died untreated. Patients with advanced stages of a disease such as breast cancer and those of a certain age are even denied treatments for certain illnesses.

Statistics from the Heritage Foundation report that in some countries with socialized medicine you are twice as likely to die from breast cancer, 2 to 3 times as likely to die from prostate cancer and, in countries like Canada, "only half as many Canadians get dialysis as Americans, per capita."[16]

Does Social Darwinism play into letting people die off when medical treatments have long been available to save them? Reports of backed-up government paperwork and bureaucratic delays granting treatment, foretell of looming nightmares for patients and their families. Will all the marches and cures for cancer and other diseases matter less or lose impetus? When governments control who gets medical treatments, is it just another form of population control born out of Social Darwinism?

Who lives? Who dies? Who decides? Will we once more view the emergence of the survival of the fittest with an unsympathetic eye towards others?

CHAPTER 12

GOD vs. DARWIN

While many scientists worship at the altar of the Living God, others do not. While many scientists worship at the altar of Darwin and evolution, others do not. Does what they believe color their views of the world regarding how people should be treated or how morality and truth should apply in our lives?

As science advances into the future, will the public be informed of the pros and cons, the positives and negatives of its potential impact on society? Will advancements be equitable among all peoples, or will it be only the wealthy and privileged that will benefit? How will the "Darwinian view" which holds that some races are superior to others weigh in on the issue?

Today, even with the latest scientific advances, scientists still do not know how life begins or how the earth, much less the universe, was formed. If science doesn't have the evolutionary evidence, wouldn't it be honest to admit it?

The origin of life continues to be a rather nasty thorn in the side of the overall evolutionary story. Many of our high school and college textbooks continue to use the 1953 Miller Urey experiment even though researchers in the field recognize that this experiment did not accurately represent early earth and therefore serves no real insight.

In his book, *Icons of Evolution: Science or Myth? Why Much of What We Teach About Evolution is Wrong*, Dr. Jonathan Wells offers the following insight:

> *So we remain profoundly ignorant of how life*
> *originated...Instead of being told the truth, we are*
> *given the misleading impression that scientists have*
> *empirically demonstrated the first step in the origin*
> *of life.*[1]

The scientific establishment should never be afraid to tell us the truth. Today, they appear to be avoiding, at all costs, the increasing logic and reason that call into doubt the evolutionary story from start to finish. Science at its most honest should not be based on some popular worldview that is as yet unsubstantiated, but on empirical data.

Russian-born, American scientist Isaac Asimov pointed out that there are some things science simply does not know: "To express all this, we can say: 'Energy can be transferred from one place to another, or transformed from one form to another, *but it can be neither created nor destroyed.*'"[2] Asimov's statement acutely begs the question: where did it all come from in the first place?

Scientists today remain perplexed by very basic, yet exceedingly complex laws of the universe that have existed since creation. Did the laws of gravity, electromagnetism, the strong and weak nuclear forces, and so on—just happen? Most believe these mysteries were put in place by God—as His creations, human beings—struggle in their limited intelligence to figure them out.

It is an undeniable fact that no scientist alive today knows how life began. No scientist alive today knows the origin of the laws and forces of nature—or the origin of the human spirit and soul. Having initially set out to find proof that would confirm evolution, many scientists have been stunned to discover that just the opposite happened. Instead of proving evolution, many found that their research resulted in an overwhelming support of creationism—which pointed to God as Creator of all.

Not surprisingly, some of America's most prestigious institutions of higher learning, among them some of our Ivy League schools, were founded on biblical principles or funded by Christian men and women. Yet, people of faith are often denigrated by many of today's so-called intellectual elitists, especially in academia. Baseless attitudes surface from time to time that people who have faith in God cannot be very intelligent, yet people of faith not only attend, but excel at these universities.

Many of our most prestigious schools that represent the highest education available in academia were started by or partially funded by Bible believing Christians. Clearly, the intellectual elites will have to take issue with Yale, Dartmouth, Columbia and Harvard—from which many of them, themselves, have graduated. Can the intellectual elitists have it both ways? Can they relegate Christianity as appealing only to the "not too bright masses," yet choose to attend schools of higher learning that were founded, funded, and also attended by the very Christians they denigrate?

Thanks to Bible believing Christians, who were also intellectuals of their day, many top universities with some of the highest educational and intellectually demanding standards in the world were able to come into existence.

So, one has to ask: Just how "intellectual" is an intellectual who chooses to follow a theory, such as the theory of evolution, without questioning its validity when there are so many cracks in its increasingly thinning veneer? Is it not being intellectually dishonest to shackle science and not allow it to pursue the truth wherever it may lead?

Study evolution—but study it honestly with flaws and all—especially if the answers impact the quality of our lives? Clearly, history shows that how we are treated and the opportunities we are afforded have long been determined by how the world views us individually and collectively.

In an effort to determine how we treat one another, the following outlines the social construct of a world that has God as Creator as compared to a world that worships at the altar of Darwin and evolution?

HOW WE TREAT ONE ANOTHER:

<u>GOD</u> and the BIBLE
vs.
<u>DARWIN</u> and EVOLUTION

<u>DARWIN</u>	<u>GOD</u>
EVOLUTION:	BIBLE:
We are distinct, separate races evolved from a mud puddle.	We are all one blood.
Some of us are less evolved and inferior.	All are created equal in the eyes of God.
The law of Naturalism allows that the weak and disabled should die out or be eliminated.	We are our brother's keeper.
Charity for the poor, the disabled, and the disenfranchised is a burden on society.	Sustain the poor, the widowed, orphaned and the disabled.
Some human traits indicate inferiority.	Racism is a sin against man and against God.
God, if He exists, does not care how we treat one another.	Love your neighbor as yourself.
Evolutionists demand that only one side of the issue be considered.	The Bible says, "Come now, and let us reason together."[3]

CHAPTER 13

CHEATING SCIENCE

I f we are to view the world we live in through a scientific microscope, shouldn't we demand that all theories, including the theory of evolution, be tested against the evidence? We should be reminded that "theories are assumptions" and assumptions are not fact. Even Darwin recognized his theory as a supposition.

Many scientists' careers rely on their pleasing the pre-established biases of their colleagues or the institutions they work for. Is it not an insult to true science for any pre-set biases to be offered or skewed as fact—when they are not yet based on evidence? The real danger is that the flaws in the theory are repeatedly being taught to our current and future generations.

Again in fairness to teachers, many are simply repeating in their classrooms what they themselves were taught in high school and college. Any quality educator will be highly disturbed that bad information is being played forward in a harmful way to their students and should demand that only honest science appear in their textbooks. It is also incumbent on parents to demand it.

So, how does the public discern whether a highly touted scientific finding, article, or reported break-through is based on fact or the evolutionist's biased agenda or prejudice? One need only look at the wording in these reports on evolution to see that most are based on mere speculations. Ask the following:

WHY IS IT THAT EVOLUTIONISTS TEND TO USE THESE WORDS?

MAY, SEEMS, COULD, INDICATES, POSSIBLY

Which one of these actual headlines, from a major cable news network's website, do you believe is true and based on fact?

1. Human Evolution **Seems To** Be Accelerating
2. Geology **May Have** Created The Perfect Cradle Of Humanity
3. Asteroid Impacts **May Have** Boosted Evolution
4. Some Neanderthals **May Have** Been Redheads
5. Skulls **Indicate** Proto-Human Males Kept Harems
6. Scientists: Biting Insects **May Have** Killed Off Dinosaurs
7. Fossilized Claw **May** Reinforce Bird-Dinosaur Link
8. Dwarf Hippo Graveyard **Could** Alter Human History
9. Human Race **May** Split In Two In Far Future
10. Missing Link Between Whales, Land Animals **Possibly** Found
11. Frozen Calf **May** Explain Mammoth's Extinction
12. Biologist: I Lost My Job Because I Don't Believe in Evolution.

Answer: Biologist: I Lost My Job Because I Don't Believe in Evolution.[1]

The only headline that is supported by actual fact is the last one. In December 2007, biologist Nathaniel Abraham filed suit in U.S. District Court stating a violation of his civil rights. Abraham charged that his employer fired him after he told a superior that he did not believe that evolution is scientific fact.

The employer insists its actions were within its rights and lawful. The employer reportedly stated essentially that it had a right to demand a belief in evolution on a project that was being supported by the U.S. Department of Health and Human Services, the primary federal agency for conducting and financing medical research. The National Institutes of Health reportedly required all scientists involved in the project to use only principles in support of evolution in their analyses and writings.[2] Per Counsel Barbara Weller, the case was dismissed on a technicality. The issue of discrimination was never addressed by the court.[3] No fault was found.

Yet, similar complaints against other employers are being reported in other work environments. Are they a bellwether of things to come? If scientific research can be skewed from the beginning by an externally imposed bias, what good is it? That then pleads the question: What "beliefs" will employers or the government be allowed to demand as job qualifications in the future? In the reverse, will employers also be able to fire you for <u>what</u> you <u>do</u> believe? Clearly, this is among the slipperiest of slopes.

But, another question to ask is why is a U.S. government agency, in charge of medical research, using taxpayers' dollars to fund a study that requires a pre-established bias towards evolution? Shouldn't science and scientists require an unbiased search for the "truth"—especially if the taxpayer is paying for it?

A major criticism is that the evolution movement has created its own industry with billions of dollars going to its supporters. Careers and organizations have been built on the back of the theory of evolution.

Dr. Jonathan Wells in his book, *Icons of Evolution,* reports that whether taxpayers agree to it or not, billions of taxpayer's dollars are being spent in an ongoing attempt to try to justify the theory of evolution.

> *In America, billions of taxpayer dollars are being spent annually without their consent to finance evolutionary origins research. The NIH (National Institute of Health) and NSF (National Science Foundation) fund research projects headed up by those already committed to an evolutionary mind-set and the*

resulting papers are used to gain professorship tenures. Students must buy evolution-oriented textbooks and attend schools and universities where only evolution is taught, presented as 'science', as an established fact.⁴

Has the industry that supports evolution gotten so big that it fears admitting it is wrong? Recall that once during a period of human history, even some scientific minded people were convinced that the earth was flat. They were wrong. People were wrong. The *Bible*, on the other hand, was right in that it stated that the earth was circular. Most, but not all, admitted the error and we moved on.

Is science being cheated by not considering all of what is possible? Is religion being penalized by those who want to discard it and approach science based on man's limited knowledge of the creation of life? Again, history may provide an answer.

One of the most acclaimed archaeologists of the nineteenth century was the avowed atheist Sir William Ramsay. A highly respected scientist and Oxford professor, Ramsay believed that the *Bible's* New Testament was not the inspired work of God, but a fraud perpetrated by men. Ramsay set out in 1881 to prove the *Bible* lacked credibility.

A skilled archaeologist, Ramsay pointedly questioned the details and documentations of places and events regarding the Apostle Paul's travels as written in the (*Bible's*) book of Acts as being unfounded. Paul's journeys were well documented by Luke, a physician and associate, who made the journey with the Apostle Paul. Dr. Luke meticulously described their travels.

For twelve years, Ramsay led an expedition in Asia Minor and Palestine, hoping to dig up the archaeological evidence to disprove the biblical accounts of Paul's journeys as written by Luke. Much to his surprise, the renowned archaeologist found Luke's accounts of the places and events in the New Testament to be stunningly accurate.

> *"Luke's history is unsurpassed in respect of its trustworthiness..."*[5] **Sir William Mitchell Ramsay**

The facts meticulously documented in the Bible astounded Ramsay.

> *"Luke is a historian of the first rank; not merely are his statements of fact trustworthy...this author should be placed among the greatest of historians."* [6]
> **Sir William Mitchell Ramsay**

Ramsay was especially impressed how small, seemingly "insignificant" geographical details were recorded exactly right in Luke's accounts. That, Ramsay claimed, was the mark of a writer who knows what he is talking about and is careful to tell everything correctly. In fact, Ramsay became so impressed with the truth of Luke's account that it led him to accept Christ.[7]

Having set out to prove that the Bible was filled with historical inaccuracies, the renowned atheist and archaeologist instead became convinced of the truth of the *Bible* in its meticulous historical detail and pronounced himself a believer of God as Creator.

Time and again, modern science has bowed to the authority and facts of the *Bible*. Biblical evidence abounds. Yet today, schools and colleges either accept or are being forced to teach evolution as the only guideline for historical scientific study. Whitman College in Walla Walla, Washington, was founded by missionaries, yet today has no religious affiliation. Whitman is an example of an educational institution where some reportedly tout Darwin and eschew creation when it comes to biological origins.

In a Whitman College press release dated September 16, 2004, the college announced it was honored to have received "...an exceedingly rare copy of Charles Darwin's biology classic, *On the Origin of Species...*" [2004] chair of Whitman's biology department, noted, *On the Origin of Species* literally changed the world overnight. Today, nothing makes sense in biology except in the light of evolution."[8]

Many top scientists across the globe, especially in the field of biology, disagree with the oft-repeated phrase that today nothing makes sense in biology except in the light of evolution. They make the case that "nothing makes sense except in the light of evidence." Kerby Anderson of Probe Ministries notes a bias embedded in evolution science today:

> *Evolutionists say you can use science to argue against the existence of God, but you can't use science in support of God.*[9]

Even Albert Einstein himself proclaims his scientific endeavors were driven by his recognition of the existence of a higher power and that his scientific principles were guided by wanting "...to know how God created this world."[10]

In their research in the lab or in nature, when scientists see the hand of an Intelligent Designer, should they disregard it because they have a predisposition that is opposed to the existence of God? The goal of real science and scientists should be to look for truth wherever they find it or wherever it leads them.

Modern View Of Race

Which begs the question: How does modern science view race? Does the color of one's skin or the shape of one's eyes matter? Many people will be surprised to know that science, today, views these characteristics as biologically trivial and as minor variations. It appears that science may be catching up to what the *Bible* has taught all along...that we are all one.

Time and again, we hear biologists say that there is no biological basis for race. Yet, people from all walks of life refuse to accept it. Is that because what is being taught in the classrooms and science labs is rooted in evolutionary theory? Is it the cultural dogma of evolution that humankind has evolved at different rates that perpetuates the perception that some of us are either superior or inferior to others?

What targets a person most, based on evolutionary theory, is skin color. It is based on a visual interpretation. But, current day biological studies reveal that all human skin starts out with the same exact substance—a pigment or coloring called *melanin*. How much of the coloring ends up in our skin, is determined by our genes. Less melanin, the lighter the skin. More melanin, the darker the skin. Since the *Bible* tells us we are all of one blood, it follows that—when God made Adam and Eve...they were, what society would call today...the ultimate example of "racial mixing". Creationists believe they are the source of all of the bloodlines of all people on earth. Skin color is just something we can easily see and identify. It's a meaningless genetic trait when it comes to assessing human value.

Racism among scientists and Darwinian evolutionists is as prevalent as it is among the general population. Racism is learned and exists globally in all people—often for the purpose of one person seeking an advantage over another. Many scientists state that our differences are cultural, not racial. Some suggest that using the word "race" to describe our minor differences should be abandoned as "meaningless," since we all come from one group, one population, one race.

Others claim that the lack of racial differentiation may prevent proper disease analysis and treatment of varied groups. One example is people who carry sickle cell anemia or the sickle cell trait. Sickle cell disease is inherited and attacks red blood cells that carry oxygen throughout the body. When the cells become misshapen and sticky, they can clog tiny vessels causing serious pain. Although the disorder can strike all races, the National Heart, Lung and Blood Institute reports that it is more common among people of African, South or Central American, Caribbean, Mediterranean, Asian Indian, Saudi Arabian and American Indian descent.[11]

Clearly, it would be incorrect to deny the research needed to combat sickle cell anemia. Ironically, few people realize that sickle cell anemia is a disease that attacks a wide spectrum of people including Greeks, Italians, Latin Americans, Asians, Africans, Arabs, African-Americans, Jews, and those from India. You can be Caucasian and have sickle cell disease or carry the trait. Medical science has traced the origination of the disease as a response to malaria. Those who only carry what is known as the sickle cell trait, are said be more resistant to malaria than most everyone else.

Under the skin, we are more alike, more homogenous than we have been led to believe. For years, science has reported that among people of differing skin color, DNA differences are tiny when compared with the widespread DNA differences that exist within a group of people who are of the same skin color. In the close of his book, *The Mismeasure of Man*, the late evolutionist Stephen J. Gould wrote:

> "But biologists have recently affirmed—as long suspected—that the overall genetic differences among human races are astonishingly small... we have found no 'race genes'."[12]

For those who hoist our differences on a medical or biological petard, it should be noted that most diseases, such as cancer, diabetes, pneumonia and yes, sickle cell, do not discriminate regardless of skin color.

CHAPTER 14

SOCIAL DARWINISM TOMORROW

W ere Darwin alive today, would he be shocked at how his theory of evolution has been applied to the devaluing of human life? As noted, Darwin admirer, British Philosopher Herbert Spencer, coined the term, "survival of the fittest", in 1851. The term became interchangeable with "Social Darwinism" as a synonym for "natural selection".

Did Darwin understand the social implications of his theory as a catalyst for racism? As a man who denied God as Creator—did he even care? Science Journalist, Denyse O'Leary, says that she was once an apologist for Darwin. But, after years of study, she now believes that Darwinism is at the root of much of society's problems.

Darwin put racism on a supposedly scientific basis. In that respect, he enabled the most virulent racism in the twentieth century. Darwin believed what he was saying in the Descent of Man...[1]

O'Leary further contends that Darwinism had a lot to do with the Holocaust because it encouraged treating people as if they were animals.

Extermination camps made much more sense when Darwinism became the creation story of the modern West. If we are all just animals, there is no longer a divine plan to flout—and therefore no final check on hatred that even a fanatic must consider.[2]

The question is: <u>Who</u> will Darwin's racists be in the future? Does humankind continue down the path paved by a flawed and unproven theory? Is evolution to be the singular mind-set taught in schools from kindergarten to universities? If so, is it science or is it indoctrination? If ideas have consequences, then who should be the arbiter of ideas taught to school children?

As scientists attempt to unravel God's mysteries of life, will we allow past and current Darwinian racial prejudices to influence how humankind will use future scientific advancements—from cloning to genetic engineering, and in future efforts, to control a burgeoning world population? Even if, as some Darwinists claim, people like Hitler and Stalin misused Darwinism, isn't it important to study those connections and how evolution was misused to such horrific degrees and not sweep it under the rug?

Clearly, scientists have long debated Darwin and his theory of evolution. They will continue the debate as long as freedom of dissent is allowed. Using his freedom of speech, former head of the Atomic Energy Commission, Dr. T.N. Tahmisian expressed the following views:

> *Scientists who go about teaching that evolution is a fact of life are great con-men, and the story they are telling may be the greatest hoax ever. In explaining evolution, we do not have one iota of fact.*[3] **Dr. T. N. Tahmisian**, former head of the Atomic Energy Commission, USA

In the past, freedom of speech has allowed scientists to speak their minds without consequence. It is a precious right that must be defended so that scientists of today and tomorrow are not penalized when their research presents opposing results to a popular view. Encroachment of such rights in the scientific community, in the classroom or society as a whole does not bode well for our future.

Consider renowned British journalist, Malcolm Muggeridge, who died in 1990. His views were in opposition to the current dominant journalistic mind-set regarding evolution. Would his career be in jeopardy if he made the following prediction today?

> *I myself am convinced that the theory of evolution, especially the extent to which it's been applied, will be one of the great jokes in the history books of the future. Posterity will marvel that so very flimsy and dubious a hypothesis could be accepted with the incredible credulity that it has.*[4] **Malcolm Muggeridge**

Some one hundred and fifty years after Charles Darwin released his grand hypothesis in his book, *On The Origin Of Species* in 1859, what has been learned? When viewed scientifically, inherent in the "150 years of Darwin" is a verifiable test—the test of time. Time has proven the racist theory wrong. Time has proven that human beings are not limited in ability and intellect based on class, disability or the color of their skin as many evolutionists have claimed.

All types of people—members of the <u>human</u> <u>race</u>—rule nations and run corporations. They are key in the arts, education, the media, and serve in the highest levels of both the government and the courts. The element of time has revealed the fallacy of Darwin's theory that among us are less evolved and, thereby, inferior human beings.

Nobel Prize winning British author, William G. Golding, best known for his novel, *Lord of the Flies*, issued this warning:

> *It was at a particular moment in the history of my own rages that I saw the Western world conditioned by the images of Marx, Darwin and Freud; and* ***Marx, Darwin and Freud are the three most crashing bores of the Western world.*** *The simplistic popularization of their ideas has thrust our world into a mental straitjacket from which we can only escape by the most anarchic violence.*[5] (emphasis added)

If the Darwinian theory has failed the test of time, then the following questions must be answered:

- Why is the <u>theory</u> of evolution being taught as <u>fact</u> in schools without challenge?
- Why is it that, <u>exposed</u> <u>evolutionary</u> <u>fraud</u> is still being printed in some school textbooks?
- Why are some U.S. government agencies using taxpayers' money to fund intentionally biased <u>pro-evolution</u> research when reportedly over two-thirds of Americans believe in God as Creator?
- Why is the National Institutes of Health allowed to target schools—grades K through 12—with "free" teachers' guides highlighting their so-called "cutting-edge" pro-evolution research?
- Why are evolutionists afraid to allow the teaching of creationism alongside evolution?
- Why are evolutionists afraid of freedom of thought?

Most secular scientists do not want to believe that they are laboring in the fields of the God who created both them and the universe they live in. Even so, these secular scientists and evolutionists admit that they still do not have the answers to how the world began or how life begins.

This recalls the often-cited comment of self-proclaimed atheist and revered evolutionist, Richard Dawkins:

> *It is absolutely safe to say that if you meet somebody who claims to not believe in evolution, that person is ignorant, stupid or insane (or wicked, but I'd rather not print that).*[6] **Richard Dawkins**

Fellow Brit, Dr Arthur E. Wilder-Smith, on the other hand, demonstrated grace and intelligence when he remained tolerant of the intolerant. Having held many distinguished positions and armed with a formidable three doctorate degrees, the first in physical organic chemistry, Dr. Wilder-Smith mastered seven languages. His positions included Full Professor of Pharmacology, University of Bergen Medical School, Norway; Full Professor of Pharmacology, University of Illinois Medical Center and received 3 Golden Apple awards. Among his writings was the book, *The Natural Sciences Know Nothing of Evolution.* (San Diego, CA: Master Books, 1981).

Dr. Wilder-Smith held that evolution did not fit with the established facts of science:

> *The Evolutionary model says that it is not necessary to assume the existence of anything, besides matter and energy, to produce life. That proposition is unscientific. We know perfectly well that if you leave matter to itself, it does not organize itself—in spite of all the efforts in recent years to prove that it does.*[7]

Also, once again appearing at odds with Dawkins is his own colleague, Dr. Stephen Jay Gould, who taught geology and paleontology at Harvard. Gould concluded that there is no *transitional evidence* anywhere in the earth's fossil record that any species has evolved to become another. Troubling to Gould was the sheer lack of physical evidence that any species has made even a *gradual* evolutionary change or that humans ever existed in a lower form than how we are today.

> *The absence of fossil evidence for intermediary stages between major transitions in organic design, indeed our inability, even in our imagination, to construct functional intermediates in many cases, has been a persistent and nagging problem for 'gradualistic' accounts of evolution.*[8] **Stephen Jay Gould**

Evolutionists are adamant in their steadfast belief in evolution though evolution has not been substantiated by scientific evidence. As modern science disassembles the Darwinian theory of evolution as not being viable, it is still being imposed in schools and laboratories. Evolutionists still boldly claim the right to set the scientific standard and charge that a belief in God is a serious detriment to anyone who wants to engage in scientific research. One of the greatest scientific minds in history—disagrees with them. In his statement that follows, Albert Einstein sought to know God's mind in that the existence of a higher power actually motivated Einstein's interest in science. One of the most renowned geniuses of all time is said to have once remarked to a young physicist:

> *I want to know how God created this world,* *I am not interested in this or that phenomenon, in the spectrum of this or that element.* **I want to know His thoughts,** *the rest are details.*[9] **Albert Einstein** (emphasis added)

According to Bryan Bissell of *Great Scientists Who Believe*:

> *Einstein built his theory of relativity on the work of three men, two of whom were Christians. The first of these Christians was Bernhard Riemann who had developed the mathematics of Riemannian Space, which Einstein found could explain the curvature of space. The other was James Clerk Maxwell whose equations and work with pre-quantum physics led directly to modern physics. Einstein's work was to some measure forced by the famous Michelson-Morely measurements of the speed of light, which showed that the speed of light is an absolute. Einstein sought and found the explanation. Edward William Morley was the Christian half of that experimental duo.*[10]

Even the undisputed brilliant mind that is still heralded by most today as the Father of Modern Science—Galileo— held an unwavering belief in God.

> *I do not feel obliged to believe that the same God who has endowed us with sense, reason, and intellect has intended us to forego their use.* [11] **Galileo**

For Galileo, biblical truth had nothing to fear from honest science. Galileo avidly applied the gift of intellect that God gave him to his scientific pursuits.

Then there is one of the most respected and brilliant minds of all of science, Sir Isaac Newton. His book *Philosophiae Naturalis Principia Mathematica* is thought to be the most influential book in the history of science. As a brilliant physicist, astronomer, mathematician, alchemist and philosopher, Newton is considered today to have laid the groundwork for scientific understanding and research from mathematics to universal gravitation to the three laws of motion of the universe, which laid the groundwork for classical mechanics. Many consider that Newton's work has had a greater impact on science than Albert Einstein. So what did this greatest of scientists have to say about a Creator God?

> *Gravity explains the motions of the planets, but it cannot explain who set the planets in motion. God governs all things and knows all that is or can be done.* **Sir Isaac Newton**[12]

Three of the greatest scientific minds in all of history spoke of their belief in God or how God motivated them in their work. Would Galileo, Isaac Newton and Albert Einstein today be held suspect? Would they be judged to be second-rate scientists by Darwinian evolutionists who consider a belief in God or a higher power to be a detriment in the research laboratory and the classroom?

What these three great minds understood is that God and science are completely compatible since God created not only the universe, but also the laws of science that hold it together and make it function. Could it be that is why Einstein sought to "know how God created this world"?

That scientists today are being forced to accept as fact a theory not yet proven—is anti-science. Yet, that is what continues to occur in laboratories and schoolrooms across the nation. Suppression by evolutionists of those who hold opposing views is not new. In 1956, leading evolutionary scientist W.R. Thompson wrote of it when he was asked to write the introduction for a new printing of Darwin's *On The Origin of Species*. Thompson obliged, but left his pro-evolution associates and Darwin supporters stunned when his introduction attacked Darwin's theory as opposed to supporting it.

> *Modern Darwinian paleontologists are obliged, just like their predecessors and like Darwin, to water down the facts with subsidiary hypotheses, which, however plausible, are in the nature of things unverifiable...This situation, where scientific men rally to the defense of a doctrine they are unable to define scientifically, much less demonstrate with scientific rigor, attempting to maintain its credit with the public by the suppression of criticism and the elimination of difficulties, is abnormal and un-desirable in science.*[13] **W. R. Thompson**

And then there are Charles Darwin's own words that expressed what he suspected all along:

> *When we descend to details, we cannot prove that a single species has changed; nor can we prove that the* supposed *changes are beneficial, which is the groundwork of the theory* [of evolution].[14] **Charles Darwin** (*of evolution* added as clarification)

Darwin's own intellect told him that his theory must be held to a higher standard should it undergo closer scrutiny and more advanced scientific review. Today, as in Darwin's day, science knows very little about how life begins or how life forms appeared. Yet, some hold fast to the theory of evolution rejecting all else. Consider the words of I.L. Cohen who was a mathematician, researcher, member of the New York Academy of Sciences, and an Officer of the Archaeological Institute of America.

> *It is not the duty of science to defend the theory of evolution, and stick by it to the bitter end, no matter*

> *which illogical and unsupported conclusions it offers.*
> *On the contrary, it is expected that scientists*
> *recognize the patently obvious impossibility of*
> *Darwin's pronouncements and predictions. Let's cut*
> *the umbilical cord that tied us down to Darwin for*
> *such a long time. It is choking us and holding us*
> *back.* **I.L. Cohen**[15]

What is clear about evolution is that *how it works* remains as muddied as the primordial pond humankind supposedly sprang from. Many evolution scientists admit they do not know how the process works due to the lack of evidence to support it. To call evolution an assumption, a belief, conjecture or a theory is valid, but it cannot and should not be called a fact. Agnostic astrophysicist Robert Jastrow, founder of NASA's Goddard Institute, summed it up this way in his often-repeated quote concerning the Big Bang when referencing the origin of the universe:

> *It is not a matter of another year, another decade of*
> *work, another measurement, or another theory; at*
> *this moment it seems as though science will never*
> *be able to raise the curtain on the mystery of*
> *creation. For the scientist who has lived by his faith*
> *in the power of reason, the story ends like a bad*
> *dream. He has scaled the mountains of ignorance,*
> *he is about to conquer the highest peak; as he pulls*
> *himself over the final rock, he is greeted by a band*
> *of theologians who have been sitting there for*
> *centuries.* **Robert Jastrow**, American astronomer,
> physicist and cosmologist, in his book *God and the*
> *Astronomers.*[16]

As evolutionists state doubts about their own assumptions, shouldn't students be allowed to study and weigh the evidence when it comes to theories—including the theory of evolution? Doesn't good science depend on it?

CHAPTER 15

RIGHTS & FREEDOMS NOW

W hat are the warning signs that a people's rights and freedoms may be in jeopardy? Again, consider Communist Russia under Stalin and Fascist Germany under Hitler. Freedom of speech first comes under government pressure. The ruling party clamps down on opposing views and holds its citizens' basic values suspect as being in opposition to what is being dictated by the state. An existing government then targets a large portion of its citizen population as being extremists and labels them "the enemy" based on their core beliefs and defense of their long held freedoms. Can it happen again?

A serious warning flag would be if a ruling political party used its power to target a large percentage of its citizens as "radical threats" or "enemies of the state" because they hold fast to and speak up for their basic freedoms, beliefs, and constitutional rights? What if then the government used one of its existing agencies to malign or propagandize against that large segment of its population? Such action would represent a serious breach of trust in the government. In America, people demand their rights as provided by the U.S. Constitution regardless of which party is in power.

What if then, as in Nazi Germany, a civilian militia is created to report on and keep the citizenry in tow? Freedom of speech and personal freedoms in general would certainly be in dire risk and such actions would be seen as a threat to freedom loving people. The press would become a tool of the state and lose its autonomy. And, as many oppressive governments have done, the citizenry would be disarmed. How does such a scenario start? It starts with disallowing opposing views wherever they exist whether in the streets, in the offices of the ACLU or in research laboratories and classrooms.

Does Social Darwinism represent a risk to our society and culture, to our freedoms? Do Darwin's racists live among us today? Will Social Darwinism breed the Darwin's racists of tomorrow? That is what we all must consider.

We have reviewed the impact of Social Darwinism on society and culture. Decisions need to be made. Do we want to live in a world that discards the morals and virtues of a Creator God in favor of a survival of the fittest philosophy whatever the costs?

And, finally, with no conclusive answers themselves, should evolutionists be allowed to appoint themselves as the "thought police" for everyone else? Clearly, a belief in God does not inhibit scientific study, but in fact, according to some of the world's greatest scientists—it motivates it. If some of the most brilliant and respected scientific minds in world history recognized the Creator God, what are evolutionists so afraid of? After all, Creation science is open to and encourages the use of all scientific research techniques as it seeks the truth of the universe.

Few people realize that in 2005, a Federal Court of Appeals ruled that Atheism, the non-belief in God, is a religion. That was in response to a prison inmate's request to conduct atheism study classes inside a Wisconsin prison. In response, the U.S. 7th Circuit Court ruled that *not believing God* is itself a religion. That same atheistic religion, that there is no Creator God, has long been cited as being at the core of Darwinian belief.

If historically, Darwinian evolution has put forth the atheistic belief that there is no Creator God, should that religion dominate all others? If atheism has been ruled a religion, then why is the religion of atheism being freely taught in public schools today under the guise of evolution? Many find it incomprehensible that while the religious belief that God is Creator is banned in the classroom, the opposing religion's belief that God is not Creator flourishes without legal intervention. At minimum, Darwinian evolution with its atheistic underpinnings should get honest scientific scrutiny, both pro and con.

Consider the comments of scientist Jonathan Wells:

> *Darwinists...routinely use public education to impose their ideas on our children. In the past few years, several states* (such as Ohio and Kansas) *adopted science curricula that included a critical analysis of evolutionary theory...Although you might think that critical analysis would be good science education, Darwinists in Ohio succeeded in banning it, and Darwinists in Kansas are now in the process of doing the same. This is not science education, but government-imposed indoctrination of our children— at our expense.*
>
> *Things are just as bad at the college level. Qualified scientists who criticize Darwinism become outcasts in their university departments, and in a growing number of cases they are losing their jobs. Taxpayer-subsidized universities—which, I think you know, are not the open-minded forums for competing ideas that they claim to be—do their best to silence all criticisms of Darwinism.*[1]

There are scientists, who after years of study and endless pursuit of elusive evolutionary fact have considered themselves duped in the process. Dr. Colin Patterson, former Senior Paleontologist at the British Museum of Natural History made the following statement at the American Museum of Natural History, New York City:

> *One of the reasons I started taking this anti-evolutionary view, was...it struck me that I had been working on this stuff for twenty years and there was not one thing I knew about it. That's quite a shock to*

*learn that one can be so misled so long...so for the
last few weeks I've tried putting a simple question to
various people and groups of people. Question is:
Can you tell me anything you know about evolution,
any one thing that is true? I tried that question on
the geology staff at the Field Museum of Natural
History and the only answer I got was silence. I
tried it on the members of the Evolutionary
Morphology Seminar in the University of Chicago, a
very prestigious body of evolutionists, and all I got
there was silence for a long time and eventually one
person said, 'I do know one thing—it ought not to be
taught in high school.'[2]*

Is one hundred and fifty years enough research, two
hundred years? When should scientists conclude that
they may be on the wrong track? When confronted with
lack of evidence to either prove or sustain a theory after
years of research that go nowhere, should other ideas be
allowed? Should scientists be allowed to pursue other
avenues of research without being penalized?

In 2005, The Wall Street Journal published an opinion
piece that raised allegations that a Smithsonian
Institution researcher was a victim of the lack of tolerance
for thoughts and ideas. In 2005, Dr. Richard Sternberg
reportedly made the grave error of allowing an article in
opposition to evolution to be printed in a biology journal.
Just by allowing an opposing opinion of ideas to be aired,
Dr. Sternberg reportedly faced the wrath of Smithsonian
officials.

Removed from his private office, a job demotion soon
followed along with allegations of ongoing harassment to
force him out as a Research Associate. The message,
according to Sternberg was—no dissenting ideas or
findings are allowed when it came to evolution. When Dr.
Sternberg took his on-the-job discrimination public, an
investigation by the House Government Reform
Subcommittee on Criminal Justice, Drug Policy and
Human Resources soon followed. In the Executive

Summary, printed in bold print in the report, the outcome of the investigation left little doubt.

> *The staff investigation has uncovered compelling evidence that Dr. Sternberg's civil and constitutional rights were violated by Smithsonian officials.*[3]

Further investigation by a subcommittee staff reported the following:

> *Officials at the Smithsonian's National Museum of Natural History created a hostile work environment intended to force Dr. Sternberg to resign his position as a Research Associate in violation of his free speech and civil rights. There is substantial, credible evidence of efforts to abuse and harass Dr. Sternberg, including punitively targeting him for investigation in order to supply a pretext for dismissing him, and applying to him regulations and restrictions not imposed on other researchers. Given the factual record, the Smithsonian's pro-forma denials of discrimination are unbelievable.* **Indeed, NMNH officials explicitly acknowledged in emails their intent to pressure Sternberg to resign because of his role in the publication of the Meyer paper and his views on evolution.**[4] (Again, bold emphasis by the writers of the Subcommittee Staff Report)

One has to ask: Did the Smithsonian NMNH officials oppose freedom of thought, the exchange of ideas and the airing of information that contradicts an as yet unproven theory—the theory of evolution? Is it part of a growing intolerance that says my opinion is the only opinion and opposing scientific research be hanged in the process?

The federal investigation stated that as a part of the intolerance, Dr. Sternberg was discredited among his colleagues, harassed by his superiors and demoted and then pressured to resign. All because he allowed another scientist's views in opposition to Darwin's theory of evolution, to be published.

During the investigation, memos and e-mails surfaced. The investigation reportedly revealed conversations among Sternberg's bosses and co-workers that questioned and denigrated his religious and political affiliations.

> *The hostility toward Dr. Sternberg at the NMNH was reinforced by anti-religious and political motivations.*[5]

The investigation revealed that one colleague went so far as to complain in an e-mail to one of Sternberg's superiors about time spent in America's "Bible Belt," by mocking that "the most fun we had was when my son refused to say the Pledge of Allegiance because of the 'under dog' part."

According to congressional investigators:

> *As a taxpayer-funded institution, such blatant discrimination against otherwise qualified individuals* based on their outside activities *raises serious free speech and civil rights concerns …Would similar expressions of disparagement be tolerated if directed at a racial minority?…Imagine a parallel situation in which government officials expressed their intent to prohibit the appointment of anyone who was found to have participated* (on their own time) *in a gay or lesbian group, or in an abortion rights group?*[7]

In the January 28, 2005, *Wall Street Journal* article, columnist and author David Klinghoffer reported the following:

> *Worries about being perceived as "religious" spread at the museum. One curator, who generally confirmed the conversation when I spoke with him, told Mr. Sternberg about a gathering where he offered a Jewish prayer for a colleague about to retire. The curator fretted: "So now they're going to think that I'm a religious person, and that's not a good thing at the museum.*[8]

The entire House Government Reform Subcommittee on Criminal Justice, Drug Policy and Human Resources investigation and report regarding the Smithsonian Institution may be read at:

> *http://www.souder.house.gov/_files/Intolerancean dthePoliticizationof ScienceattheSmithsonian.pdf*

Again, consider renowned astrophysicist, Dr. Hugh Ross. Once a non-believer, his research and studies took him to the outer limits of the universe. It is there, at the outer reaches of the cosmos, where he says he discovered the handiwork of a supreme Creator. After years of advanced scientific study and research, Dr. Ross asserts that:

"Faith and science are not enemies, they are allies."[9]

If, as Ross says, that Creation and science are allies, shouldn't serious scientists have the freedom to research the evidence whether they are a believer or a non-believer? Such a freedom to seek the truth is after all, not only a basic tenet of science, but also of American life.

American history shows that the American people dissent when freedom of expression, freedom of ideas, freedom of opposing views, freedom of religious and political affiliations come under fire. They are basic freedoms upon which America was built.

Any scientific theory that has been shown by history to breed racial discrimination, work place discrimination, hatred and death, must be examined more closely? The theory of evolution must be held up to scrutiny. Do not the future rights and freedoms of our citizens depend on not repeating the mistakes of the past?

Keep in mind the words of Dr J.Y. Chen, Chinese Paleontologist:

In China it's OK to criticize Darwin, but not the government, but in the United States it's OK to criticize the government but not Darwin.[10]

If basic rights and freedoms can reportedly come under threat by officials with a scientific bent for Darwin and his theory of evolution at a tax-dollar funded, government-run institution, then what further lengths will Darwinian supporters go to suppress opposing views and castigate others? Many doubt the validity of evolution, even many evolutionists themselves. Doubting evolution should not be a punishable offense in the workplace, classroom, laboratory or society. The goal is to teach honest, unbiased and unadulterated science. There is no need to simultaneously teach that there is no God. Science and faith are compatible. To think otherwise indicates a separate agenda all its own.

I.L. Cohen's credentials read: Mathematician, Researcher, Member of the New York Academy of Sciences, and Officer of the Archaeological Institute of America. Years ago, Cohen spoke up. Today, hundreds of scientists refuse to remain silent. They are challenging the theory of evolution as if future generations depend on it.

Every single concept advanced by the theory of evolution (and amended thereafter) is imaginary as it is not supported by the scientifically established facts of microbiology, fossils, and mathematical probability concepts. Darwin was wrong. The theory of evolution may be the worst mistake made in science.[11] **I.L. Cohen**

The following abridged statement should be considered and carefully read regarding an often-repeated quote that appeared in the *New York Review of Books* in 1997. Famed Harvard geneticist Richard Lewontin let the cat out of the bag in what is considered to be one the most honest statements made by an evolutionist.

We take the side of science in spite of the patent absurdity of some of its constructs, in spite of its failure to fulfill many of its extravagant promises of health and life, in spite of the tolerance of the scientific community for unsubstantiated just-so stories, because we have a prior commitment, a commitment to materialism...Moreover, that materialism is absolute, for we cannot allow a Divine Foot in the door.[12]

Demand Freedom Now

Freedom is the foundation of the American way of life. Freedom of thought is a God given right. The teaching of God as Creator was a constitutional right. No one, no government, at any level, should be allowed to take that away from a truly free people.

DEMAND YOUR FREEDOM NOW.

In the final quote, Charles Darwin acknowledged his own doubts about his Theory of Evolution meeting scientific criteria. He had the following to say about this unproven theory that, in a relatively short period of time, has changed humankind's view of itself:

"I am quite conscious that my speculations run quite beyond the bounds of true science."
Charles Darwin[13]

www.DarwinsRacists.com

Part IV

NOTES

Part I—Yesterday: Charles Darwin—The Man At The Center

1. Charles Darwin, *Descent of Man, and Selection In Relation To Sex*, Introduction, John Murray Publishers, London, 1871. From Wikisource, Descent of Man/Chapter I—Wikisource. Retrieved from http://en.wikipedia.org/wiki/Descent_of_Man_1 on January 15, 2009.
2. Charles Darwin, *Descent of Man, and Selection In Relation To Sex*, John Murray Publishers, London, Chapter One, pg. 1, 1871. Source: WikiSource, Retrieved from http://en.wikisource.org/wiki/Descent_of_Man/Chapter_I on September 21, 2007.
3. Wikipedia, the free encyclopedia, Charles Darwin photo (Public Domain). Retrieved from http://en.wikipedia.org/wiki/File:Charles_Darwin.jpg on January 14, 2009.
4. Darwin, Charles, *The Descent of Man, and Selection in Relation to Sex, Princeton*, NJ: Princeton University Press, 1981 (1871), 1:201.
5. Charles Darwin, The Descent of Man, 2nd edition, New York, A L. Burt Co., 1874, p. 178.

Chapter 1—Evolution

1. William Dembski, quotes Jonathan Wells, "Wells vs. Shermer at Cato Institute" (October 14, 2006), Uncommon Descent: Serving the Intelligent Design Community. Retrieved from http://www.uncommondescent.com/evolution/wells-vs-shermer-at-cato-institute/ on October 7, 2008.
2. Jonathan Wells comments, Darwin's Deadly Legacy, DVD, Jerry Newcombe and John Rabe, Coral Ridge Ministries Media, (D. James Kennedy, Host.) 2006–2007 (www.CoralRidge.org).

(See also: Jonathan Wells, *Icons of Evolution: Science or Myth*, Regnery Publishing, Inc., Washington, D.C, October, 2000.)

3. Richard Dawkins, "Put Your Money on Evolution," *The New York Times* (April 9, 1989) section VII p. 35. Retrieved from http://www.mercatornet.com/articles/view/evolution_speci al_issuedarwinism_strikes_back/, January 17, 2009. (Also see: April 9, 1989 byline/review of book: *Blueprints: Solving the Mystery of Evolution*, (by Maitland A. Edey and Donald C. Johanson), *The New York Times*. Retrieved from http://www.simonyi.ox.ac.uk/dawkins/WorldOfDawkins-archive/Dawkins/Work/Reviews/1989–04–09review_blueprint.shtml on January 15, 2009. Comment was also then quoted from Josh Gilder, a creationist, in his critical review, "PBS's 'Evolution' series is propaganda, not science" (September, 2001). Additional source: D. James Kennedy (Host), <u>Darwin's Deadly Legacy</u>, Jerry Newcombe and John Rabe, DVD, Coral Ridge Ministries Media, Inc., 2006–2007, (www.CoralRidge.org).

4. Sean D. Pitman (compiled by), "Thoughts on Evolution From Scientists and Other Intellectuals," p. 10 (quotes Phillip E. Johnson from PBS documentary "In the Beginning: The Creationist Controversy," May 1995, by Randall Balmer (host/writer/interviewer), produced by WTTW, Chicago, IL.) Retrieved from http://www.detectingdesign.com/quotesfromscientists.html on January 15, 2009—which was retrieved from original website (by Sean D. Pitman—compiler) http://naturalselection.0catch.com/Files/quotesfromscienti sts.html on January 15, 2009.

5. Richard Dawkins, *The Blind Watchmaker: Why the Evidence of Evolution Reveals a Universe Without Design*, 1986, page 6. New York: W.W. Norton. EvoWiki. Retrieved from http://cc.msnscache.com/cache.aspx?q=darwin+made+it+p ossible+to+be+an+intellectually+fulfilled+atheist&d=753149 17933755&mkt=en-US&setlang=en-US&w=e97d7b9,d66fb20a on January 15, 2009. Retrieved from http://bevets.com/evolution.htm on July 2, 2008.

6. Editor/Compiler, <u>An Atheist Fairy Tale</u>, "Evolution and Atheism"—quotes Richard Dawkins. (Site content provides various quotes.) Retrieved on http://bevets.com/evolution.htm on July 2, 2008.

7. Sean D. Pitman (compiled by), "Thoughts on Evolution From Scientists and Other Intellectuals," (quotes William B. Provine, Professor of Biological Sciences, Cornell University,

"Darwin Day" website, University of Tennessee Knoxville, 1998), p. 10. Retrieved from http://www.detectingdesign.com/quotesfromscientists.html on January 15, 2009 which was retrieved from original website (by Sean D. Pitman) http://naturalselection.0catch.com/Files/quotesfromscientists.html on January 15, 2009.

8. John MacArthur, The MacArthur New Testament Commentary, Acts 13–28, Moody Press/Chicago, 1996, p. 51.

9. Jonathan Sarfati, "Review of *Climbing Mount Improbable"* by Richard Dawkins (Penguin Books Ltd, Harmondsworth, Middlesex, England), A Book review of Dawkins Climbing Mount Improbable—Journal of Creation (TJ) 12(1):29–34, April 1998, Creation Journal, Creation Ministries International, CreationOnTheWeb.com (www.creationontheweb.com). Retrieved from http://www.creationontheweb.com/content/view/1855 on July 1, 2008.

10. Editor/Staff, <u>Creation or Evolution—Does It Really Matter What You Believe</u>, "Scientists, Creation and Evolution," quotes Antony Flew, Editors/Compilers, Good News Magazine, United Church of God. Retrieved on http://www.gnmagazine.org/booklets/EV/scientistscreation.htm on December 18, 2008. (Also see; *The Sunday Times*, December 12, 2004.)

11. Gary R. Haberman, Interviews Anthony Flew: <u>My Pilgrimage from Atheism to Theism</u>, for the Evangelical Philosophical Society's journal "Philosophia Christi"—Craig J. Hazen, Editor, Winter issue, 2005. Retrieved from http://www.biola.edu/antonyflew/flew-interview.pdf December 18, 2008.

12. Stephen Goode, Interviews Phillip E. Johnson, "Johnson Challenges Advocates of Evolution," Insight on the News, October 25,1999; and http://www.geocities.com/truedino/pjohnson.htm, 1999 New World Communication, Inc.; (reprinted with permission of Insight). Retrieved from http://www.creationontheweb.com/content/view/1855 on January 17, 2009.

13. I.L. Cohen, *Darwin Was Wrong: A Study in Probabilities*, New York: NW Research Publications, Inc., 1984, pp. 214–215. (See also: Vance Ferrell, *Science VS. Evolution*, Chapter 23a: Scientists Speak, quotes from I.L. Cohen, *Darwin Was Wrong: A Study in Probabilities*. Retrieved from http://www.pathlights.com/ce_encyclopedia/sci-ev/sci_vs_ev_23.htm on September 22, 2008.)

Chapter 2—Creation

1. American Religious Identification Survey (ARIS), March 9, 2009. The Graduate Center of the City University of New York.
2. "Louis Pasteur." From Conservapedia. Quotes Louis Pasteur. Retrieved from http://www.conservapedia.com/Louis_Pasteur on March 4, 2009.
3. Ibid.
4. Editors/Compilers Creation-Evolution Encyclopedia, *Evolution Cruncher*, Scientists Speak, Chapter 23, "Science VS. Evolution: Chapter 23a, 3—Scientists Speak Against Evolution," p. 5. Quotes James Gorman, *"The Tortoise or the Hare?" Discover*, October, 1980. p. 88. Retrieved from http://www.pathlights.com/ce_encyclopedia/sci-ev/sci_vs_ev_23.htm on September 3, 2008.
5. Genesis 1:26—"Scripture taken from the HOLY BIBLE. NEW INTERNATIONAL VERSION. Copyright © 1973, 1978, 1984 International Bible Society. Used by permission of Zondervan Bible Publishers."
6. Genesis 1:27—"Scripture taken from the NEW AMERICAN STANDARD BIBLE®, Copyright © 1960, 1962, 1963, 1968, 1971, 1971, 1972, 1975, 1977, 1995 by The Lockman Foundation. Used by permission."
7. Ibid, Genesis 2:7
8. Edmund J. Ambrose (Professor of Cell Biology at the University of London), *The Nature and Origin of the Biological World,* 1982, John Wiley & Sons, pg. 164.
9. Acts 17:26—"Scripture taken from the NEW AMERICAN STANDARD BIBLE®, Copyright © 1960, 1962, 1963, 1968, 1971, 1971, 1972, 1975, 1977, 1995 by The Lockman Foundation. Used by permission."
10. Peter Hammond, "William Wilberforce—Setting the Captives Free," Frontline Fellowship. This article comes from *The Greatest Century of Missions* by Peter Hammond. Retrieved from http://www.frontline.org.za/articles/settingcaptives_free.htm on July 3, 2008.
11. Ibid.
12. D.M.S. Watson, "Adaptation," in Nature, Vol. 123 [sic Vol. 124], 1929, p. 233.

Chapter 3—Charles Darwin's Family & Friends

1. Adrian Desmond and James Moore, *Darwin: The Life of a Tormented Evolutionist*, New York, New York: Warner Books, 1991, p. 239.
2. Charles Darwin, *The Descent of Man, and Selection in Relation to Sex*, Princeton, NJ: Princeton University Press, 1981 (1871), 1:201.
3. Richard Weikart, *From Darwin to Hitler: Evolutionary Ethics, Eugenics, and Racism in Germany*, New York, NY: Palgrave Macmillan, 2004, p. 106.
4. Ashton Nichols, "Romanticism & Ecology, The Loves of Plants and Animals: Romantic Science and the Pleasures of Nature," at Dickinson College re: Erasmus Darwin, *Zoonomia: (Temple of Nature*, IV, 419–28), 1794. (Orrin Wang, Series Editor, Romantic Circles (University of Maryland), "Romantic Circles Praxis Series.) Retrieved from http://www.rc.umd.edu/praxis/ecology/nichols/nichols.html on July 3, 2008.
5. Gerardus D. Bouw, Ph.D., "A Brief Introduction To The History of Evolution," Brief History of Evolution, www.geocentricity.com. Retrieved from http://www.geocentricity.com/ba1/no85/evolhist.html on January 20, 2009.
6. Charles Darwin's Notebooks, pg. 268, Handlist of Darwin Papers at the University Library, Cambridge University Press, 1960. Additional source: Barrett, Paul H. (Transcribed and Annotated), *Early Writings of Charles Darwin*, Book Two, commentary on M and N Notebooks by Howard E. Gruber. Retrieved from The Complete Works of Charles Darwin Online, 2002–8, Director John van Wyhe. Retrieved from http://darwin-online.org.uk/content/frameset?viewtype=text&itemID=F1582&pageseq=1 on July 5, 2008.
7. Francis Darwin, ed., *The Life of Charles Darwin*, (John Murray, London, 1902); "Autobiography" originally published in 1887. Additional source: The Victorian Web, Charles Darwin, Autobiography, Chapter II, page 5. Retrieved from http://www.victorianweb.org/science/darwin/darwin_autobiography.html on January 19, 2009.
8. Charles Darwin's Autobiography (edited by Sir Francis Darwin), Henry Schuman, New York, 1950, p. 21. Quoted by Russell Grigg, "Darwinism: It was all in the family. Erasmus Darwin's famous grandson learned early about evolution,"

First published: Creation 26(1):16–18, December, 2003. (www.answersingenesis.org) Retrieved from http://www.answersingenesis.org/creation/v26/i1/darwinism.asp on July 5, 2008.

9. Charles Darwin, *A Naturalist's Voyage Around the World*, 1st Edition, May, 1860, Chapter 10, Tierra del Fuego, p.225, The Project Gutenberg Etext—The Voyage of the Beagle, by Charles Darwin Project Gutenberg Etext A Naturalist's Voyage Round the World by Darwin #18 in our series by Charles Darwin. Retrieved from http://www.darwinsgalapagos.com/Darwin_voyage_beagle/darwin_beagle_chapter_10.html on July 6, 2008.

10. Ken Ham, *Darwin thought natives were advanced animals*, Answers In Genesis, First published: Creation 16 (1): 37, 1993. (www.answersingenesis.org) Retrieved from http://www.answersingenesis.org/creation/v16/i1/darwin.asp on August 22, 2007.

11. Charles Darwin, The Voyage of the Beagle (1845) in *So Simple a Beginning: The Four Great Books of Charles Darwin*, New York, New York: W.W. Norton & Company, Ltd., 2006, p. 327.

12. Ibid.

13. Charles Darwin, Voyage of the Beagle (1839), pp. 330–331.

14. Jonathan Weiner, *The Beak of the Finch: A Story of Evolution in Our Time*, New York, NY: Alfred A. Knopf, Inc., 1994, p. 9.

15. Peter Brent, *Charles Darwin: A Man of Enlarged Curiosity*, Harper and Row, New York, 1981, p. 221. (Additional sources: Andrew J. Bradbury, "Charles Darwin—The Truth?" (Part 8, Going Public—Maybe, Theory, Fact or Fiction?). Retrieved from http://www.bradburyac.mistral.co.uk/dar8.html on January 19, 2009.

16. Herbert Spencer in his Principles of Biology of 1864, vol. 1, p. 444.

17. Harun Yahya (aka: Adnan Oktar), "Darwin's Racism and Colonialism," Alliance In The Intellectual Struggle, Union of Faiths.com. From Charles Darwin, *The Descent of Man*, 2nd edition, New York, A L. Burt Co., 1874, p. 171. Retrieved from http://www.unionoffaiths.com/article5_5.html on October 8, 2008.

18. Erasmus Alvey Darwin. Retrieved from http://en.wikipedia.org/wiki/Erasmus_Alvey_Darwin on September 3, 2008.

19. Sean D. Pitman, compiler, October, 2004, "Thoughts on Evolution From Scientists and Other Intellectuals," The

Philosophy of Evolution, p. 2, Erasmus Alvey Darwin, in a letter to his brother Charles, after reading his new book, "The Origin of Species," in Darwin, F., ed., "The Life of Charles Darwin," [1902], Senate: London, 1995, reprint, p. 215. Retrieved from http://www.detectingdesign.com/quotesfromscientists.html on September 3, 2008.

20. "Adam Sedgwick." Wikipedia—the free encyclopedia, Letter to Charles Darwin from Adam Sedgwick, November 24th, 1859, in *The Correspondence of Charles Darwin* vol. 7, pg. 396. Retrieved from http://en.wikipedia.org/wiki/Adam/Sedgwick on October 8, 2008.

21. Harun Yahya (aka Adnan Oktar), *The Disasters Darwin Brought To Humanity* (Chapter 2, p. 1), from Prof. Adam Sedgwick per A.E. Wilder-Smith, Man's Origin Man's Destiny, The World for Today Publishing, 1993, p. 166. Retrieved from http://www.harunyahya.com/disasters03.php on October 8, 2008.

22. Ibid. (http://www.harunyahya.com/disasters03.php)

23. Aldous Huxley, *Confessions of a Professed Atheist*, Report: Perspective on the News, Vol. 3, June, 1966, p. 19. (Other sources: Aldous Huxley: *Ends and Means: An Inquiry Into The Nature Of Ideals and Into Methods Employed For Their Realization*, Harper, New York, 1st ed., p.270. Quoted by Sean D. Pitman, compiler, "Thoughts on Evolution From Scientists and Other Intellectuals," October, 2004, p. 11. Retrieved from http://www.detectingdesign.com/quotesfromscientists.html on September 26, 2008.

24. Editor/Compilers, Creation-Evolution Encyclopedia (evolutionfacts.org), Evolution Cruncher Chapter 23, "Science VS Evolution: Chapter 23a Scientists Speak: 1-Evolutionists Explain Their Objective," p. 2, Quotes Aldous Huxley, *Confessions of a Professed Atheist*, Report: Perspective on the News, Vol. 3, June, 1966, p. 19. Retrieved from http://www.pathlights.com/ce_encyclopedia/sci-ev/sci_vs_ev_23.htm on October 26, 2008.

25. "Charles Darwin's Illnesses." From Wikipedia—the free encyclopedia. Retrieved from http://en.wikipedia.org/wiki/Charles_Darwin%27s_illness On January 2, 2008. Darwin Correspondence Project—Letter 4834—Darwin, C. R. to Chapman, John, May 16, 1865. Retrieved from

http://www.darwinproject.ac.uk/darwinletters/calendar/en
try-4834.html on March 16, 2008.

26. Carolyn Douglas, "Changing Theories of Darwin's Illness,"
article published on Purdue University Website, May 24,
1990, Prof. Gene Joy—(Source: http://omni.cc.purdue.
edu/~sbenning/el102c/Darwin.html). Retrieved from
http://www.christianforums.net/viewtopic.php?t=16429 on
July 11. 2008. Note original source: Edward L. Rempf
(1918), "Charles Darwin—The Affective Sources of His
Inspiration and Anxiety-Neurosis," *Psychological Review* 5.
Retrieved from http://en.wikipedia.org/wiki/Charles_
Darwin%27s_illness on January 2, 2008.

27. "Charles Darwin's Illness." From Wikipedia—the free
encyclopedia. Retrieved from
http://en.wikipedia.org/wiki/Charles_Darwin%27s_illness
on January 2, 2008. (Additional source: Ibid, Rankine Good,
The Life of Shawl, Lancet, January 9, 1954, pgs. 106–107.
Retrieved from http:// www.christianforums.net/viewopic.
php?t=16429 on January 2, 2008.)

28. Carolyn Douglas, "Changing Theories of Darwin's Illness,"
article published on Purdue University Website, (Prof. Gene
Joy), May 24, 1990—regarding tropical disease expert Dr.
A.W. Woodruff's comments on Darwin's illnesses and
dismissing chagas. (As sourced from http://omni.cc.purdue.
edu/~sbenning/el102c/Darwin.html). Retrieved from
http://creationwiki.org/Darwin%27s_Illness on January 21
2009.

29. Sean D. Pitman (compiler), October, 2004, "Thoughts on
Evolution From Scientists and Other Intellectuals" (p. 2):
The Philosophy of Evolution: quote from H. J. Lipson, F.R.S.
(Professor of Physics, University of Manchester Institute of
Science and Technology, UK), "A physicist looks at evolution"
Physics Bulletin, vol. 31, 1980, Retrieved from
http://www.detectingdesign.com/quotesfromscientists.html
on September 26, 2008.

Chapter Four—Eugenics Nightmare

1. John M. Brentnall and Russell M. Grigg, "Darwin's slippery
slide into unbelief," First published: Creation 18(1):34–37,
December, 1995: references Gertrude Himmelfarb, *Darwin
and the Darwinian Revolution,* Chatto and Windus, London,
1959, pp. 10, 318—from Life and Letters of Charles Darwin,
D. Appleton and Co., New York, 1911, Vol. 1, p. 318.
(www.answersingenesis.org) Retrieved from

http://www.answersingenesis.org/creation/v18/i1/slide.asp on July 8, 2008.

2. "Francis Galton" Public Domain Photo: Retrieved from http://en.wikipedia.org/wiki/image:Francis_Galton_1850s.jpg on January, 2009. Scanned from Karl Pearson's The Life, Letters and Labors of Francis Galton.

3. John Cavanaugh-O'Keefe, *The Roots of Racism and Abortion: An Exploration of Eugenics*, www.eugenics-watch.com, Chapter Two: Francis Galton and the Eugenics Society (quotes Francis Galton). Retrieved from Eugenics Watch website http://www.eugenics-watch.com/roots/chap02.html on July 8, 2008. (Original source: F. Galton, Memories of My Life (London, 1908). Karl Pearson, The Life, Letters, and Labours of Francis Galton, Cambridge University Press (London, 1914–30).

4. "Eugenics." From Wikipedia, the free encyclopedia. Retrieved from http://en.wikipedia.org/wiki/Eugenics on July 8, 2008.

5. "Leonard Darwin." From Wikipedia, the free encyclopedia. Retrieved from http://en.wikipedia.org/wiki/Leonard_Darwin on January 21, 2009.

6. Charles Darwin, *On The Origin of Species*, 6th ed. NYU, 1988, p. 154.

7. Nancy Pearcey in her 2004 book, *Total Truth: Liberating Christianity From Its Cultural Captivity*, reports that Holmes experiences in the Civil War turned him against the Christian faith of his college years and came under the influence of Herbert Spencer, one of Darwin's early disciples in England. Pearcey writes on page 229 that "From then on he began to argue that evolution applies not only to physical organisms, but also to the sphere of beliefs and convictions.

8. Oliver Wendell Holmes, Jr., Buck v. Bell, 274 U.S. 200 (1927). Retrieved from http://en.wikipedia.org/wiki/Oliver_Holmes%2C_Jr on July 8, 2008. (Public Domain photo (author unknown) retrieved from http://en.wikipedia.org/File:Oliver_Wendell_Holmes ,_1902.jpg.)

9. John Hawkins, *Right Wing News*, Interviewing Jonah Goldberg. Retrieved from http://www.rightwingnews.com/mt331/2008/01/interviewing_jonah_goldberg_ab.php on October 23, 2008.

10. George Grant comments, Darwin's Deadly Legacy, DVD, Jerry Newcombe and John Rabe, TV broadcast and DVD, Coral Ridge Ministries Media, Inc., (D. James Kennedy

(Host), 2006–2007 (www.CoralRidge.org). Dr. George Grant, *Grand Illusions: The Legacy of Planned Parenthood*, Cumberland House Publishing, Fourth Edition, pg. 59, 1999; Margaret Sanger, *The Pivot of Civilization*, 1922, Brentanos Publishers, New York.

11. Margaret Sanger photo, Wikipedia—the free encyclopedia. Author: Underwood & Underwood. Retrieved from http://en.wikipedia.org/wiki/File:MargaretSanger-Underwood.LOC.jpg on March 24, 2009.

12. Margaret Sanger quotes, Eads Home Ministries (Editor) quotes Margaret Sanger (*The Woman Rebel*, Volume I, Number 1. Reprinted in *Woman and the New Race*. New York: Brentanos Publishers, 1922; Retrieved from http://www.eadshome.com/MargaretSanger.htm on January 22, 2008.

13. George Grant, Commentary.Net, Margaret Sanger. Retrieved from http://www.commentary.net/commentary/s80p978.htm on October 27, 2008.

14. Edwin Black, *War Against The Weak: Eugenics and America's Campaign To Create A Master Race* (New York: Thunder's Mouth Press, 2004), 127. Source: Quote references prepared by David Noebel and posted on Worldview Times, August 2, 2006 (Brannon Howse, President and Founder, www.worldviewtimes.com). Retrieved from http://www.christianworldviewnetwork.com/article.php/939/David_Noebel October 27, 2008.

15. History of Indigenous Australians, Wikipedia, the free encyclopedia from: Margaret Sanger, *What Every Girl Should Know*, pg. 47, Springfield, Ill.: United Sales Co., 1920. Retrieved from http://en.wikipedia.org/wiki/History_of_Indigenous_Australians on January 22, 2009.

16. Margaret Sanger, April 1932 *Birth Control Review*.

17. "Immigration to the United States," Wikipedia, the free encyclopedia, Puck magazine (USA) Oct 3, 1888. (Public Domain) Retrieved from http://en.wikipedia.org/wiki/American_immigration#Demography on January 24, 2009.

18. Raimundo Rojas, (Director Hispanic Outreach), "PPFA Tries to Put a Religious Gloss on Abortion." Retrieved from http://www.nrlc.org/news/2004/NRL04/ppfa_tries_to_put_a_religious_gl.htm on January 23, 2009. Margaret Sanger letter to Dr. Clarence Gamble, 1939, Sanger/Sophia Smith Collection quoted in Linda Gordon, *Woman's Body, Woman's Right: Birth Control in America*, New York: Grossman

Publishers, 1976, Second edition, New York: Penguin Books, 1990, 332–33). Retrieved from http://www.nrlc.org/bal/sanger.html on July 7, 2008.

19. Clenard H. Childress, Jr. BlackGenocide.org, "The Negro Project: Margaret Sanger's EUGENIC Plan for Black Americans," p. 4, quotes Dorothy Ferebee, Retrieved from http://blackgenocide.org/negro04.html on July 10, 2008.

20. Ethan Clive Osgoode, Evolution: Darwinism-Eugenics v0.4.8, An Investigation Into Inbred Science, Retrieved from http://www.freerepublic.com/~ethancliveosgoode/ on July 15, 2008.

21. George Grant, *Grand Illusions: The Legacy of Planned Parenthood*, pg. 59, Cumberland House Publishing, Fourth Edition, 2000 retrieved from "The Human Right of Family Planning," IPPF London, 1984 and "A Strategy for Legal Change," IPPF London, 1981.

22. "Margaret Sanger," Wikipedia—the free encyclopedia, p. 6. quoted from: Margaret Sanger, "What Every Boy and Girl Should Know," 1915, pg. 140. Retrieved from http://en.wikipedia.org/wiki/Margaret_Sanger on January 22, 2009.

23. Keith Garvin, ABC News, State Secret: Thousands Secretly Sterilized, ABC News' Keith Garvin originally report for "*World news Tonight*" on April 23, 2005. Retrieved from http://abcnews.go.com/WNT/Health/story?id=708780 on July 16, 2008.

24. Tony Riddick comments, Darwin's Deadly Legacy, DVD, Jerry Newcombe and John Rabe, TV broadcast and DVD, Coral Ridge Ministries Media, Inc., (D. James Kennedy (Host), 2006–2007 (www.CoralRidge.org).

25. George Grant comments, Darwin's Deadly Legacy, DVD, Jerry Newcombe and John Rabe, TV broadcast and DVD, Coral Ridge Ministries Media, Inc., (D. James Kennedy (Host), 2006–2007 (www.CoralRidge.org).

26. Alveda King quoted in "Dr. Alveda King, MLK's Niece, Calls Planned Parenthood Racist" by CottShop, Sacred Scoop, April 17, 2008, Christian World News. Retrieved from http://sacredscoop.com/?p=799 on July 17, 2008.

27. Editor/Staff, Dr. Alveda C. King, "Bio Sketch Alveda King," www.priestsforlife.org. Retrieved from http://www.priestsforlife.org/staff/alvedaking.htm on July 16, 2008.

28. Martin Luther, King, Jr. speech at the Lincoln Memorial, Washington, D.C., August, 1963. Retrieved from http://www.martinlutherkingjrarchive.com/ on September

3, 2008. Staff, The Seattle Times website, "Martin Luther, King, Jr. and the Civil Rights Movement," Special Report, Newspapers in Education. Retrieved from http://seattletimes.nwsource.com/special/mlk/ on September 3, 2008.

29. Amanda Galiano, About.com—Littlerock Newsletter, "Rags to Riches: Johnny Cash," Article. Retrieved from http://littlerock.about.com/cs/artsentertainment/a/johnny cash.htm on September 3, 2008.

30. Jesse Lee Peterson (Reverend), *Abortion: Black Genocide*, Commentary on WorldNet Daily, February 13, 2008. Retrieved from http://www.wnd.com/index.php?pageId= 56202 on September 3, 2008.

31. Jesse Lee Peterson, Reverend Jesse Lee Peterson comments to author, Bond Action, Inc., www.Bondaction.org, April 2, 2009.

Chapter 5—Darwin's Racists

1. Harun Yahya: An Invitation to Truth, *The Evolution Deceit*, The Refutation of Darwinism—introduction. Retrieved from http://www.hyahya.org/evolution_introduction.php on July 11, 2008.

2. Ibid.

3. Editor/Staff, Creation Tips, "Why evolution breeds monsters like Hitler, Trotsky and Stalin," Staff article. Retrieved from http://www.users.onaustralia.com.au/rdoolan/tyrants.htm l; on July 11, 2008.

4. E. Yaroslavsky, *Landmarks in the Life of Stalin*, Foreign Languages Publishing House, Moscow, 1940, pp. 8–9. Source: Creation Tips, "Why evolution creates tyrants like Hitler, Stalin and Trotsky," Staff article. Retrieved from http://www.users.onaustralia.com.au/rdoolan/tyrants.htm l; on July 11, 2008.

5. Edward E. Ericson, Jr., "Solzhenitsyn—Voice from the Gulag," *Eternity*, October 1985, pp. 23, 24 as quoted by Harun Yahya (a.k.a. Adnan Oktar), *The Disasters Darwinism Brought To Humanity*, Chapter, 4. Retrieved from http://www.harunyahya.com/disasters05.php on January 27, 2009.

6. Paul G. Humber, *Stalin's Brutal Faith*, October, 1987 article, Institute for Creation Research from: Edward E. Ericson, Jr., "Solzhenitsyn—Voice from the Gulag," *Eternity*, October 1985, pp. 23. Retrieved from http://www.icr.org/article/276/ on January 27, 2009. Full

190

text of incident may be found: Aleksander I. Solzhenitsyn, *The Gulag Archipelago* (New York: Harper & Row, 1973), p. 7.

7. Michael Berenbaum, *The World Must Know*, The United States Holocaust memorial Museum, pg. 125ff. Retrieved from "The Holocaust"— http://en.wikipedia.org/wiki/The_Holocaust on July 13, 2008 references Donald L. Niewyk, *The Columbia Guide to the Holocaust*, Columbia University Press, 2000, p.45; *Encyclopedia Britannica*, 2007.

8. Adolf Hitler, *Mein Kampf*, (1925) quoted by D. James Kennedy (Host), Darwin's Deadly Legacy, DVD, Jerry Newcombe and John, Coral Ridge Ministries, 2006–2007 (www.CoralRidge.org).

9. John Hawkins, *Right Wing News*, Interviews author Jonah Goldberg about his book *Liberal Fascism: The Secret History of the American Left, From Mussolini to the Politics of Meaning*. Retrieved from http://www.rightwingnews.com/mt331/2008/01/interviewing_jonah_goldberg_ab.php on October 23, 2008.

10. Phillip E. Johnson quote, Staff, Discovery Institute, *Praise For "From Darwin To Hitler,"* August 1, 2004, Center for Science and Culture, Retrieved from http://www.discovery.org/a/2173 on January 28, 2009.

11. Charles Darwin, *The Life and Letters of Charles Darwin*, Vol. 2, p. 270 as quoted by Richard Weikart, *From Darwin to Hitler: Evolutionary Ethics, Eugenics, and Racism in Germany*, by Richard Weikart, Palgrave Macmillan, p. 324.

12. Richard Weikart comments, Darwin's Deadly Legacy, DVD, Jerry Newcombe and John Rabe, TV broadcast and DVD, Coral Ridge Ministries Media, Inc., (D. James Kennedy (Host), 2006–2007 (www.CoralRidge.org). Weikart is author of *From Darwin to Hitler, Evolutionary Ethics, Eugenics and Racism in Germany*. Palgrave MacMillan, New York, NY, 2004.

13. Frederick Nietzsche from Wikipedia—the free encyclopedia. Retrieved from http://en.wikipedia.org/wiki/Nietzche on February 25, 2008. (Public Domain)

14. D. James Kennedy, *The Coral Ridge Hour*, pre-recorded, Channel 10, (KTTV, Los Angeles. CA), repeated on June 22, 2008.

15. Adolf Hitler, From Wikipedia—the free encyclopedia, File:Drawing of Adolf Hitler.jpg. Retrieved from http://en.wikipedia.org/wiki/File:Drawing_of_Adolf_Hitler.jpg on March 24, 2009

16. Charles Darwin, From Wikipedia—the free encyclopedia, File:Charles Darwin 01.jpg, photo by Julia Margaret Cameron, 1869. Retrieved from http://en.wikipedia.org/wiki/File:Charles_Darwin_01.jpg on March 24, 2009.

17. Charles Darwin, *On The Origin Of Species By Means of Natural Selection, or The Preservation of Favoured Races in the Struggle for Life, November 22, 1859,* John Murray Publishers, London, England. See also: Editor/Staff, The Talk Origins Archive, The Origin of Species—Chapter 4: Natural Selection by Charles Darwin. (www.talkorigins.org) Retrieved from http://www.talkorigins.org/faqs/origin/chapter4.html on April 9, 2009.

18. Richard Weikart comments, Darwin's Deadly Legacy, Jerry Newcombe and John Rabe, TV broadcast and DVD, Coral Ridge Ministries Media, Inc., (D. James Kennedy (Host), 2006–2007 (www.CoralRidge.org). Weikart is author of *From Darwin to Hitler, Evolutionary Ethics, Eugenics and Racism in Germany.* Palgrave MacMillan, New York, NY, 2004.

19. "Arthur Keith" from Wikipedia—the free encyclopedia, quotes Arthur Keith, 1946. Evolution and Ethics, G. P. Putnam 's Sons, New York, p. 230. Retrieved from http://en.wikipedia.org/wiki/Arthur_Keith, on January 28, 2009.

20. Richard Weikart, *From Darwin to Hitler: Evolutionary Ethics, Eugenics, and Racism in Germany,* Palgrave MacMillan, New York, NY, 2004. (Additional source: Darwin's Deadly Legacy, Jerry Newcombe and John Rabe, TV broadcast and DVD, Coral Ridge Ministries Media, Inc., (D. James Kennedy (Host), 2006–2007. (www.CoralRidge.org).

21. Jonathan Sarfati, "Nazis planned to exterminate Christianity," Creation Archive, Volume 24, Issue 3 22(1):4, 1999). (www.answersingenesis.org) Retrieved from http://www.answersingenesis.org/creation/v24/i3/nazi.asp on September 22, 2007.

22. Editor/Staff: Adherents, "The Religious Affiliation of Adolf Hitler—German Dictator, Nazi Leader" (p. 2) quotes from Jadwiga Biskupska, "Hitler & Triumph of the Will: A Nazi Religion in the Catholic Style" in *Undergraduate Quarterly,* Cornell University, September/November 2004, page 147, (accessed from http://www.undergradquarterly.com/EJournal/2004Q2/Biskupska.pdf). Retrieved from

192

http://www.adherents.com/people/ph/Adolf_Hitler.html on
March 24, 2009.

23. Jonathan Sarfati, "The Holocaust and evolution" (Guest
Editorial), Creation Archive, Volume 22, Issue 1, p.1 quotes
Julius Streicher. References: *Nuremberg Trial Proceedings*
Volume 2, The Avalon Project at the Yale Law School (see:
www.yale.edu/lawweb/avalon/imt/proc/11–21–45.htm.)
(www.answersingenesis.org) Retrieved from http://
www.answersingenesis.org/creation/v22/i1/holocaust.asp
on July 14, 2008.

24. Ibid. p. 2. (Source: see also http://www-camlaw.rutgers.edu/
publications/law-religion/nuremberg.html.)

25. *Nuremberg Trial Proceedings* Volume 2, The Avalon Project at
the Yale Law School,
www.yale.edu/lawweb/avalon/imt/proc/11–21–45.htm.
(www.answersingenesis.org) Retrieved from
http://www.answersingenesis.org/creation/v22/i1/holocau
st.asp on September 22, 2007. (www.answersingenesis.org)

26. Denyse O'Leary, The ID Report, Post Details:—Part Three:
"Darwin and the Holocaust," February 20, 2008. Retrieved
from
http://www.arn.org/blogs/index.php/2/2008/02/20/part_t
hree_darwin_and_the_holocaust on July 14, 2008.

27. Denyse O'Leary, The ID Report, Post Details: Part Two,
"Darwin and scientific racism," 2–20–08. Retrieved from
http://www.arn.org/blogs/index.php/2/2008/02/20/part_t
wo_darwin_and_scientific_racism on July 14, 2008.

28. Sunayana Sadarangani, Editor, "An Inspiration to fight
back!"—Famous people with cerebral palsy, Children With
Cerebral Palsy, Indian Child, 2000. Retrieved from
http://www.indianchild.com/CerebralPalsy/famous-people-
with-cerebral-palsy.htm on September 3, 2008.

29. "Anthony Romero." From Wikipedia, the free encyclopedia.
Retrieved from
http://en.wikipedia.org/wiki/Anthony_Romero on
September 30, 2008.

30. Phillip E. Johnson quote, Staff, Discovery Institute, *Praise
For "From Darwin To Hitler,"* August 1, 2004, Center for
Science and Culture, Retrieved from
http://www.discovery.org/a/2173 on January 28, 2009.

Chapter 6—Frauds, Fakes & Mistakes

1. Jerry Newcombe and John Rabe, D. James Kennedy (Host), *Darwin's Deadly Legacy*, DVD, Coral Ridge Ministries, 2006–2007. (www.coralridge.org)
2. Randall Niles, "Problems With The Fossil Record," All About The Journey. Retrieved from http://www.allaboutthejourney.org/problems-with-the-fossil-record.htm on September 8, 2008.
3. Randall Niles, "Problems With The Fossil Record," All About The Journey, quotes Colin Patterson, personal communication (April, 1979) to Luther Sunderland, *Darwin's Enigma: Fossils and Other Problems,* New Leaf Publishing Group, Master Books, Green Forest, AR, 4th edition, 1988, pp. 88–90. Retrieved from http://www.allaboutthejourney.org/problems-with-the-fossil-record.htm on September 8, 2008.
4. Editor/Staff, WhistleBlower Magazine, "Evolution fraud in current biology textbooks: Exposed as fakes decades ago, publishers still include them," WorldNetDaily, Article p. 1, July 7, 2001, Retrieved from http://www.worldnetdaily.com/news/article.asp?ARTICLE_ID=23532 on January 30, 2009. (Quote from *Haeckel's Frauds and Forgeries* by authors J. Assmuth and Ernest R. Hull, 1915.)
5. Russell Grigg, "Fraud rediscovered—It has long been known that one of the most effective popularizers of evolution fudged some drawings, but only now has the breathtaking extent of his deceit been revealed," Creation Magazine Volume 20, Issue 2, pages 49–51, March 1998 quotes Dr. Michael Richardson as interviewed by Nigel Hawkes, The Times (London) p. 14, August 11, 1997. (www.answersingenesis.org) Retrieved from http://www.answersingenesis.org/creation/v20/i2/fraud.asp on January 30, 2009. (Additional reference: Embryonic fraud lives on, New Scientist 155(2098):23, September 1997 and Michael Richardson et al, Anatomy and Embryology Journal, *Science* and *New Scientist* 196(2): 91–106, 1997.)
6. Editor/Staff, WhistleBlower Magazine, "Evolution fraud in current biology textbooks: Exposed as fakes decades ago, publishers still include them," WorldNetDaily, Article p. 2, July 7, 2001, Retrieved from http://www.worldnetdaily.com/news/article.asp?ARTICLE_ID=23532 on January 30, 2009.

7. Douglas Futuyma, quoted in *Icons of Evolution*, Jonathan Wells, Regnery Publishing, Inc., Washington, DC, 2000, p.108.

8. Stephen J. Gould, quoted in *Icons of Evolution*, Jonathan Wells, Regnery Publishing, Inc., Washington, DC, 2000, p. 109

9. "Ernst Haeckel Biography" at biographybase, Retrieved from http://www.biographybase.com/biography/Haeckel_Ernst.html on August 27, 2008.

10. Jonathan Wells, *Icons of Evolution*, Regnery Publishing, Inc., Washington, DC, 2000,pp. 85–90.

11. Royal Truman, "What biology textbooks never told you about evolution," TJ Archive, Vol. 15, Issue 2, August, 2001—A review of *Icons of Evolution: Science or Myth? Why much of what we teach about evolution is wrong* by Jonathan Wells, Regnery Publishing Inc., 1st edition, Washington, D.C., 2000. (www.answersingenesis.org) Retrieved from http://www.answersingenesis.org/tj/v15/i2/textbooks.asp; on January 28, 2008

12. Charles Blinderman, *The Piltdown Inquest*, Chapter 15, pp 235–239, Prometheus Books, 700 East Amherst Street, Buffalo, New York, 1986. Quotes Gertrude Himmelfarb, *The Darwinian Revolution* (1959). Retrieved from fhttp://www.clarku.edu/~piltdown/The_Piltdown_Inquest/chapters/chapter15.html on August 29, 2008.

13. Dust jacket review *Icons of Evolution: Science or Myth?*, author Jonathan Wells, Regnery Publishing, Inc., 1st edition, Washington, D.C., October, 1, 2000.

14. Amedee Forestier—Artist, <u>Nebraska man illustration of two humanlike creatures</u>, drawing by Amedee Forestier for the Illustrated London News, (1922), File:NebraskaMan.jpg— Wikipedia, the free encyclopedia. Retrieved from http://en.wikipedia.org/wiki/File:NebraskaMan.jpg on January 28, 2009.

15. Hank Hanegraaff (source), *The Face That Demonstrates The Farce Of Evolution*, Word Publishing, Nashville, 1998, pp.50–52). Additional Source: Copyright 1994 by the Christian Research Institute, P.O. Box 500-TC, San Juan Capistrano, CA 92693.

16. en.Wikipedia.org File:Huxley—Man's Place in Nature.jpg, (1863). Retrieved from http://en.wikipedia.org/wiki/File:Huxley_-_Mans_Place_in_Nature.jpg on January 22, 2009.

17. Time Magazine, Science, ed.: source: "Upgrading Neanderthal Man," Time Magazine, May 17, 1971, Vol. 97,

No. 20). Retrieved from
http://www.time.com/time/magazine/article/0,9171,94438
0,00.html on January 31, 2009.

18. *"Ass Taken for Man,"* London Daily Telegraph, May 14, 1984.
Chris Ashcroft, Staff, Northwest Creation Network, "Evolution
Fraud," Evolution from the Creation Perspective, Network,
Retrieved from http://www.nwcreation.net/evolutionfraud.html
on January 31, 2009. (source: "Skull fragment may not be
human," Knoxville News-Sentinel, 1983.)

19. Marvin Lubenow, *Bones of Contention: A Creationist Assessment
of the Human Fossils,* Baker Books, Grand Rapids, MI,
December, 1992, pp. 115–134.

20. 20. Big Daddy?—"Copyright 2000 by Jack T. Chick.
Reproduced by permission of Chick Publications. Website:
www.chick.com/reading/tracts/0055/0055_01.asp"

21. David Menton comments, <u>Darwin's Deadly Legacy</u>, Jerry
Newcombe and John Rabe, DVD, Hosted by Dr. D. James
Kennedy, Coral Ridge Ministries Media, 2006–2007
(www.CoralRidge.org).

22. David Catchpoole, "New evidence: Lucy was a knuckle-
walker," answersingeneisis.org, May 5, 2000.
(www.answersingenesis.org) Retrieved from
http://www.answersingenesis.org/docs2/4256news5-5-
2000.asp on August 29, 2008.

23. Paul Abramson, Editor, "12 Quotes from Leading
Evolutionists," www.creationism.org. Referring to comments
made by Richard Leakey (Director of National Museums of
Kenya) in The Weekend Australian, May 7–8, 1983,
Magazine, p. 3. *The Revised Quote Book,* edited by Dr. A.
Snelling, PhD, pub. by: Creation Science Foundation,
Australia. Retrieved from
http://www.creationism.org/articles/quotes.htm on
February 14, 2008.

24. Charles Blinderman, *The Piltdown Inquest,* Chapter 15, pp
235–239, Prometheus Books, 700 East Amherst Street,
Buffalo, New York, 1986. Quotes Dr. Gary E. Parker of the
Institute of Creative Research (1981). Retrieved from
http://www.clarku.edu/~piltdown/The_Piltdown_Inquest/c
hapters/chapter15.html on August 29, 2008. (also:
http://www.clarku.edu/~piltdown/The_Piltdown_Inquest/T
PI-MAP.html.)

25. Randall Niles, "Problems With The Fossil Record," (Quote
source: David B. Kitts, *Evolution,* vol. 28, p. 467.) *What
Happened To Me?: Reflections of a Journey,*
AllAboutTheJourney.org. Retrieved from

196

http://www.allaboutthejourney.org/problems-with-the-fossil-record.htm on September 8, 2008.

26. Niles Eldredge, "Niles Eldredge: Evolutionist, Biography, 2005–2006. Retrieved from http://nileseldredge.com/biography.htm on September 15, 2008.

27. Niles Eldredge, *Time Frames: The Rethinking of Darwinian Evolution and the Theory of Punctuated Equilibrium*, Simon & Shuster: NY, NY, 1985, p. 144). Retrieved from http://english.sdaglobal.org/research/qotcratn.htm on September 5, 2008.

28. Ryan N. Helms, <u>Evolution is not a proven theory</u>, Washington County News, Holmes County Times-Adviser, February 12, 2008. Retrieved from http://www.chipleypaper.com/news/evolution_1463__article.html/darwin_theory.html on October 8, 2008. Quotes: Charles Darwin, *On The Origin Of Species By Means of Natural Selection, or The Preservation of <u>Favoured</u> Races in the Struggle for Life*, John Murray Publisher, London, England, November 22, 1859, p.143.

29. Charles Darwin, *On the Origin of Species by Means of Natural Selection, or the Preservation of Favoured Races in the Struggle for Life*, John Murray Publisher, London, England, 1859, p. 162.

30. Michael J. Behe, *Darwin's Black Box: The Biochemical Challenge to Evolution*, Free Press, New York,1996 as quoted from MSU SACS "Quotes from Famous Scientists," Mississippi State University Online. Retrieved from http://www.msstate.edu/org/sacs/quotes.html on February 4, 2009.

31. Jonathan Wells, <u>An Evaluation of Ten Recent Biology Textbooks: And their selected use of evolution</u>, (evaluated by Dr. Jonathan Wells), Discovery Institute, A Report for the Center for the Renewal of Science and Culture, 2000. Retrieved from http://www.arn.org/docs/wells/textreport900.pdf on February 4, 2009.

32. Jonathan Wells, *Icons of Evolution: Science or Myth?*, Regnery Publishing, Inc., 1st edition, Washington, D.C., October, 1, 2000. Press Release on release of book. Retrieved from http://www.sedin.org/bol/bol007B_2000.html on October 7, 2008.

33. Steven Jay Gould, *Ontogeny and Phylogeny*, Belknap-Harvard Press, Cambridge, Massachusetts, 1977, pp. 127–128.

34. Stephen Jay Gould, "Will We Figure Out How Life Began?" April 10, 2000, Time Magazine, 155 (14): 92–93. Retrieved from http://www.time.com/time/magazine/article/0,9171,99661 9,00.html on August 31, 2008.

35. Compiler/Staff, Jim Crow Museum of Racist Memorabilia at Ferris State University—"Question of the Month: Human Zoos," October, 2006. Retrieved from http://www.ferris.edu/jimcrow/question/oct06.htm on February 4, 2009.

36. en.Wikipedia.org File:Ota Benga at Bronx Zoo.jpg. Source: http://www.npr.org/programs/atc/features/2006/09/ota_b enga/bronx_lg.jpg. Retrieved from http://en.wikipedia.org/wiki/File:Ota_Benga_at_Bronx_Zoo. jpg on February 5, 2009.

37. "Ota Benga." From Wikipedia, the free encyclopedia, p. 1. Retrieved from http://en.wikipedia.org/wiki/Ota_Benga on October 30, 2008.

38. Geoffrey C. Ward, "The Man in the Zoo," American Heritage Magazine, The Life and Times, (Quotes: Reverend Gordon comments from 1906.), Volume 43, Issue 6, October, 1992. Retrieved from http://www.americanheritage.com/articles/magazine/ah/1 992/6/1992_6_12.shtml on February 5, 2009.

39. *The New York Times*, "Man And Monkey Show Disapproved By Clergy," September 10, 1906. Retrieved from http://query.nytimes.com/mem/archive-free/pdf?_r=1&res=9C04E7D81F3EE733A25753C1A96F9C9 46797D6CF onFebruary 5, 2009.

40. "Ota Benga." From Wikipedia, the free encyclopedia, p. 2. Retrieved from http://en.wikipedia.org/wiki/Ota_Benga on October 30, 2008.

41. Ken Ham comments, Darwin's Deadly Legacy, Jerry Newcombe and John, (Host Dr. D. James Kennedy). Coral Ridge Ministries Media, DVD, 2006–2007 (www.CoralRidge.org).

42. David Monaghan, 'The body-snatchers,' The Bulletin (Australian weekly), November 12, 1991, pp. 30–38. (The article states that journalist Monaghan spent 18 months researching this subject in London, culminating in a television documentary called *Darwin's Body-Snatchers*, which was aired in Britain on October 8, 1990.) "Darwin's bodysnatchers: new horrors—People deliberately killed to provide 'specimens' for evolutionary research" by

Carl Wieland, Answers in Genesis, Brisbane, Australia. See: Darwin's Bodysnatchers,' *Creation* 14(2):16–18, March, 1992. (Carl Wieland 'Darwin's Bodysnatchers,' *Creation*, 12(3):21, June-August, 1990.) Retrieved from http://www.creationontheweb.com/content/view/1067/ on September 22, 2007.

43. David Monaghan, 'The body-snatchers,' *The Bulletin*, November 12, 1991, pp. 30–38. See also: Harun Yahya (aka Adnan Oktar*), Social Darwinism And The Favored Races Myth,* The Holocaust Violence.com, Retrieved from http://holocaustviolence.com/holocaust_3.html on February 6, 2009.

44. David Monaghan, 'The body-snatchers,' <u>The Bulletin</u> (Australian weekly), November 12, 1991, p. 33. (Monaghan is quoting Dr. Rae Sumner, a lecturer at the Queensland Institute of Technology's School of Language and Literacy Education.) "Darwin's bodysnatchers: new horrors—People deliberately killed to provide 'specimens' for evolutionary research" by Carl Wieland, Answers in Genesis, Brisbane, Australia. See also: Darwin's Bodysnatchers,' *Creation* 14(2):16–18, March, 1992. (Carl Wieland 'Darwin's Bodysnatchers,' *Creation*, 12(3):21, June-August, 1990.) Retrieved from http://www.creationontheweb.com/content/view/1067/ on September 22, 2007.

45. Staff/Editor, "Focus: news of interest about creation and evolution," answersingenesis.org, Creation, 18(1):7–9, December, 1995. Retrieved from http://www.answersingenesis.org/creation/v18/i1/focus.asp on September 2, 2008.

46. "Racism in Africa" Source: Wikipedia, the free encyclopedia. Retrieved from http://en.wikipedia.org/wiki/Racism_in_Africa on March 9, 2008.

47. Tom Jones, "Exiles of the Kalahari," Mother Jones, January/February, 2005 Issue. Retrieved from http://www.motherjones.com/news/dispatch/2005/01/01_800.html.on September 6, 2008.

48. Miriam Ross, "UN condemns Botswana government over Bushmen eviction," Survival—The Movement for Tribal Peoples website, News from Survival International, March 13, 2006. Retrieved from http://www.survival-international.org/news/1454 on September 2, 2008. See also: "UN Condemns Botswana's Racism," August 30, 2002.

Retrieved from http://www.survival-international.org/news/81 on February 6, 2009.

Chapter 7—The Thought Police

1. Niles Eldredge, "Confessions of a Darwinist," *The Virginia Quarterly Review*: p. 15, Issue: Spring, 2006. Retrieved from http://www.vqronline.org/articles/2006/spring/eldredge-confessions-darwinist/#fn11 on September 15, 2008.
2. John Alroy, <u>Asa Gray (1810–1888)</u>, Lefalophodon, An Informal History of Evolutionary Biology Web Site. Retrieved from http://www.nceas.ucsb.edu/~alroy/lefa/Gray.html on October 1, 2008.
3. Alan Keyes, "Endowed By Their Creator," Renew America website, October 7, 2007. Retrieved from http://www. renewamerica.us/columns/keyes/071007 on October 1, 2008.
4. Sean D. Pitman, "Thoughts on Evolution From Scientists and Other Intellectuals," compiled, October, 2004, quoting Boyce Rensberger, *How the World Works*, William Morrow, NY, 1986, pp. 17–18. Rensberger is an ardently anti-creationist science writer. Retrieved from http://www.detectingdesign.com/quotesfromscientists on September 26, 2008.
5. Anthony Romero comments on <u>NOW—PBS</u>, interviewed by Bill Moyers (Host), Public Affairs Television, Transcript, December 17, 2004. Retrieved from http://www.pbs.org/now/printable/transcript351_full_print .html on September 30, 2008.
6. Ian Taylor comments, <u>Darwin's Deadly Legacy</u>, Jerry Newcombe and John, (Host Dr. D. James Kennedy), Coral Ridge Media Ministries, DVD, 2006/2007 (www.CoralRidge.org).
7. Al Sharpton, *Al On America*, by Rev. Al Sharpton and Karen Hunter (Contributor), Kensington, October 1, 2002, p. 89. Retrieved from http://www.ontheissues.org/2008/Al_Sharpton_ Education.htm#School_Choice on September 10, 2008.
8. Witold Walczak comments, *Redeeming Darwin: Discovering the Designer*, DVD, Probe Ministries and EvanTell, 2007.
9. Bharat Rathode, "Teaching Intelligent Design violates First Amendment, says Federal Judge," EARTHtimes.org, December 22, 2005. Retrieved from

http://www.earthtimes.org/articles/show/4757.html on
February 14, 2008.

10. David K. DeWolf, John G. West, Casey Luskin and Jonathan
Witt, *Traipsing Into Evolution: Intelligent design and the
Kitzmiller v. Dover Decision*, Discovery Institute Press,
Seattle, Washington, 2006, p. 123.

11. Stephen Meyer comments, Evolution Vs. God in the
Classroom—Interview on The Big Story with John Gibson,
May 9, 2005, FoxNews.com, Gibson and Nauert, Transcript
(eMediaMillWorks, Inc. 2005). Retrieved from
http://www.foxnews.com/story/0,2933,155943,00.htm on
February 7, 2009.

12. George Grant comments, Coral Ridge Media Ministries,
Darwin's Deadly Legacy, Jerry Newcombe and John Rabe,
(Host Dr. D. James Kennedy).
DVD, 2006/2007 (www.CoralRidge.org).

13. Alan Keyes on Evolution, Republican presidential candidate
Alan Keyes' remarks on evolution were made at Hylton High
school in Virginia on 2/27/00. Skeptic Mag Hotline
Newsletter excerpts from March 6, 2000
(http://www.skeptic.com/) as compiled by Keith Devens on
KeithDevens.com, March 7, 2008. Retrieved from
http://keithdevens.com/weblog/archive/2000/Mar/07/Ala
n-Keys.evolution on October 1, 2008.

14. Emerson Thomas McMullen, The Implications of the
Cambrian Explosion for Evolution (text), Georgia Southern
University Professor quotes Stephen Jay Gould, *Wonderful
Life, The Burgess Shale and the Nature of History*, New York:
W. W. Norton Co., 1989, p. 208. Retrieved from
http://personal.georgiasouthern.edu/~etmcmull/CAM.htm
on February 9, 2009.

15. Jonathan Wells on the Cambrian Explosion and Darwinism
as a Science–Stopper posted by Gil Dodgen, Uncommon
Descent, Q&A Part 3, January 14, 2007. Retrieved from
http://www.uncommondescent.com/intelligent-design/qa-
part-3-jonathan-wells-on-the-cambrian-explosion-and-
darwinism-as-a-science-stopper/ on November 5, 2008.

16. J. Madeleine Nash, "When Life Exploded," quotes Samuel
Bowring of MIT, Time Magazine, December 4, 1995, p. 70.
Also see: Article online at http://www.time.com/time/
magazine/article/ 0,9171,983789–4,00.html.

17. Matthew Vanhorn, The Eye of the Trilobite, article retrieved
from Http://www.apologeticspress.org/articles/2021 on
March 23, 2009. Additional source: Frank Sherwin and
Mark Armitage, Trilobites,—The Eyes Have It! Retrieved from

http://www.trueorigin.org/trilobites_eyes.asp on March 23, 2009.

18. Editor/Staff, "Scientists, Creation and Evolution," The Good News Magazine, <u>Creation or Evolution: Does It Really Matter What You Believe?</u>, International Church of Christ, quotes Sir Ernst Chain. Retrieved from http://www.gnmagazine.org/booklets/EV/scientistscreation .htm on September 19, 2008.

19. Hugh Ross Biography, <u>About Our Founder</u>, Reasons To Believe website: www.reasonstobelieve.org. (www.reasons.org) Retrieved from http://www.reasons.org/about/staff/ross.shtml?main on October 4, 2008

20. Editor/Staff, Creation, Dinosaurs and the Flood—Facts, "Kansas Decision Rhetoric." Comments that Stephen Hawking...has stated that he is looking for a better model. Retrieved from http://www.sixdaycreation.com/facts/ creation/general/jan2000.html on October 25, 2008.

21. Isaiah 55:8–9, "Scripture taken from the NEW AMERICAN STANDARD BIBLE®, Copyright © 1960, 1962, 1963, 1968, 1971, 1971, 1972, 1975, 1977, 1995 by The Lockman Foundation. Used by permission. (Zondervan Publishers)

22. Ibid. Genesis 1:1—modified. For original verse see NAS.

23. Albert Einstein photo—File:Einstein 1921 by F Schmutzer 4.jpg, Wikipedia, the free encyclopedia. Retrieved from http://en.wikipedia.org/wiki/File:Einstein1921_by_F_Schm utzer_4.jpg on March 10, 2009.

24. David Calhoun (compiler), "Christian Apologetics Quotes Scientists and Philosophers," <u>Dave's Main Gate</u>, October 23, 2007. (Quotes Albert Einstein from *Ideas And Opinions* by Albert; Carl Seelig, Ed.; Sonya Bargmann, trans., Einstein (Author), Bonanza Books, 1954. Retrieved from http://themaingate.net/apologetics/quotes.php on February 10, 2009. Also: Albert Einstein quotes from *Science, Philosophy and Religion*, A Symposium, published by the Conference on Science, Philosophy and Religion in Their Relation to the Democratic Way of Life, Inc., New York, 1941. (Source: Albert Einstein on Religion and Science.) Retrieved from http://www.sacred-texts.com/aor/einstein/einsci.htm#TWO on February 10, 2009.

25. Einstein: Science and Religion, *Short Comments on Einstein's Faith*, Quoted in *The New York Times* obituary April 19, 1955, Albert Einstein, Alice Calaprice, ed., *The Quotable Einstein*, Princeton University Press, New jersey,

September 16, 1996, pp. 145–161, (New edition is available under *The New Quotable Einstein.*) Retrieved from http://www.einsteinandreligion.com/faithcomments.html on October 7, 2008.

26. Isaiah 40:22—"Scripture taken from the NEW AMERICAN STANDARD BIBLE®, Copyright © 1960, 1962, 1963, 1968, 1971, 1971, 1972, 1975, 1977, 1995 by The Lockman Foundation. Used by permission. (Zondervan Bible Publishers)

27. View of the Earth seen by the Apollo 17 crew traveling toward the moon, National Space Science Data Center (NSSDC Photo Gallery Earth) and NASA Photo, December 7, 1972. Retrieved from http://nssdc.gsfc.nasa.gov/photo_gallery/photogallery-earth.html on March 24, 2009.

28. Romans 1:20, "Scripture taken from the HOLY BIBLE. NEW INTERNATIONAL VERSION. Copyright © 1973, 1978, 1984 International Bible Society. Used by permission of Zondervan Bible Publishers."

29. Tom Carpenter, "Christians must be careful not to elevate science beyond its place," (Part 2), Creation Science Defense, January 16, 2001, pp.47–48. Retrieved from http://www.creationdefense.org/48.htm on December 18, 2008.

30. Editor/Compiler, Creationism Resources (and Intelligent Design, Too), Quotes segment, p. 2. Retrieved from http://peregrin.jmu.edu/~johns2ja/evolution/Creation/cre ation.htm on February 11, 2009.

31. Bryan Bissell. Great Scientists Who Believe, SDA Global (www.english.sdaglobal.org), quotes Wernher von Braun's foreword to book: *From Goo to You by Way of the Zoo* by Harold Hill (Plainfield, New Jersey: Logos International, 1976), p. xi. Retrieved from http://english.sdaglobal.org/research/sctstbel.htm on September 4, 2008 and from MSU SACS—Quotes from Famous Scientists retrieved from http://www.msstate.edu/org/sacs/quotes.html on December 18, 2008.

32. Jerry Bergman, "Arno A. Penzias: Astrophysicist, Nobel Laureate," American Scientific Affiliation, Astronomy/Cosmology Page. P.1, excerpt from Browne, Malcolm. "Clues to the Universe's Origin Expected." *The New York Times*, Mar. 12, 1978, p. 1, col. 54. Retrieved from http://www.asa3.org/ASA/PSCF/1994/PSCF9-94Bergman.html on December 19, 2008.

33. Editor/Compiler, <u>Quotes from Famous Scientists</u>, MSU SACS, Mississippi State University, Retrieved from http://www.msstate.edu/org/sacs/quotes.html on December 18, 2008.

34. Phillip R. Johnson, "How To Meet the Evils of the Age," The Spurgeon Archive, Chapter 4, quotes Charles H. Spurgeon from <u>The Sword and the Trowel.</u> Retrieved from http://www.spurgeon.org/misc/aarm04.htm on October 7, 2008.

35. Ibid, pp. 4–5.

36. Jonathan Wells, *Icons of Evolution: Science or Myth?,* Regnery Publishing, Inc., Washington, DC, 2000.

37. William Dembski, quote from Jonathan Wells, "Wells vs. Shermer at Cato Institute" (October 14, 2006), Uncommon Descent: Serving the Intelligent Design Community. Retrieved from http://www.uncommondescent.com/evolution/wells-vs-shermer-at-cato-institute/ on October 7, 2008.

38. Ray Bohlin, *Redeeming Darwin:Discovering the Designer,* DVD, Probe Ministries and EvanTell, 2007.

39. Roger Highfield, "Do our genes reveal the hand of God?," Telegraph.co.uk, March 26, 2003. Retrieved from http://www.telegraph.co.uk/connected/main.jhtml?xml=%2Fconnected%2F2003%2F03%2F19%2Fecfgod19.xml on September 13, 2008.

40. Bryan Bissell, <u>Great Scientists Who Believe</u>, SDA Global (www.english.sdaglobal.org), p.43, quotes Francis Crick, *Life Itself: Its Origin and Nature,* W.W. Norton, New York, 1982, p.88. Retrieved from http://english.sdaglobal.org/research/sctstbel.htm#_Toc71912181 on September 5, 2008.

41. Randall Niles, *Miracle Of Life,* All About the Journey, AllAbouttheJourney.org. Quotes Francis Crick, *Life Itself—its origins and Nature,* Futura, 1982. Retrieved from http://www.allaboutthejourney.org/miracle-of-life on September 14, 2008.

42. "Alexander Oparin (redirected from Aleksandr Oparin)," Wikipedia, the free encyclopedia. Source: <u>Great Soviet Encyclopedia</u>, 3rd edition, entry on "Опарин." Retrieved from http://en.wikipedia.org/wiki/Alexander_Oparin on February 12, 2009.

204

Chapter 8—Part II: Today—A Beacon Of Freedom

1. Harun Yahya (aka Adnan Oktar), "How the Theory of Evolution Breaks Down in the Light of Modern Science," Darwinism Refuted.com, topic: Irreducible Complexity. Source: Charles Darwin, *The Origin of Species: A Facsimile of the First Edition*, Harvard University Press, 1964, p. 189. Retrieved from http://darwinismrefuted.com/irreducible_complexity.html on October 10, 2008.

2. Bryan Bissell, "Great Scientists Who Believe," SDA Global (www.english.sdaglobal.org), p. 46, quotes Stephen Jay Gould's October, 1983 speech: As reported in "John Lofton's Journal"-The Washington Times, February 8,1984. Retrieved from http://english.sdaglobal.org/research/sctstbel.htm#_Toc71 912181 on September 5, 2008.

3. Luther Sunderland, *Darwin's Enigma—Ebbing the Tide of Naturalism*, New Leaf Publishing Group, Green Forest Arkansas, 1988, Chapter 5, p.7, (www.creationism.org). Retrieved from http://www.creationism.org/books/ sunderland/DarwinsEnigma/DarwinsEnigma_05MoreProble ms.htm on February 13, 2009. See also: Stephen Jay Gould, "Is a New and General Theory Emerging?"—Lecture at Hobart and William Smith Colleges, Feb. 14, 1988.

4. John Saboe, Evolution Is Dead, quotes Stephen Jay Gould, *The Panda's Thumb*," (Book of essays from his column "The View of Life," in Natural History Magazine), W.W. Norton & Company, New York, 1980 regarding "Rarity of Transitional Forms," pp.179–181. Retrieved from http://english.sdaglobal.org/research/qotcratn.htm on October 10, 2008.

5. Guillermo Gonzalez and Jay W. Richards, *The Privileged Planet: How Our Place in the Cosmos Is Designed for Discovery*, Washington, DC, Regnery Publishing, Inc., 2004, p. 444 and The Privileged Planet: The Search for Purpose in the Universe, www.illustramedia.com, Illustra Media , 2004, (Film/DVD) 60 minutes. See also: http://expelledthemovie.com

6. Staff/Discovery Institute, "Dr. Guillermo Gonzalez And Academic Persecution," updated February 8, 2008. Retrieved from http://www.discovery.org/scripts/viewDB/index.php?comm and=view&id=2939 on March 30, 2009. (Also: http://www.discovery.org/a2939) Additional source: Denyse

O'Leary, "Astronomer Guillermo Gonzalez's appeal denied."—
Academic freedom petition launched, The ID Report.
Retrieved from
http://www.arn.org/blogs/index.php/2/2008/02/07/astro
nomer_guillermo_gonzalez_s_appeal_d and http:post-
darwinist.blogspot.com/ on March 28, 2009.

7. Daniel Lapin (Rabbi) comments. "One Nation Under God,"
    Jerry Newcombe, John Rabe and D. James Kennedy (Host),
    DVD, Coral Ridge Ministries, 2005. (www.CoralRidge.org)

8. "Thomas Jefferson." From Wikipedia—the free encyclopedia,
    quotes Letter to Danbury Baptist Association, CT, January
    1, 1802. Retrieved from http://en.wikipedia.org/wiki/
    ThomasJefferson on October 14, 2008.

9. Donald Lutz (Professor) comments to author, Houston, TX,
    May 2008.

10. Donald Lutz comments, D. James Kennedy, Jerry
    Newcombe and John Rabe, One Nation Under God, DVD,
    Coral Ridge Ministries Media, 2005, hosted by Dr. D. James
    Kennedy (www.CoralRidge.org).

11. Abraham Lincoln. Photo, From Wikipedia, the free
    encyclopedia, File:Abraham Lincoln head on shoulders
    photo portrait.jpg. Retrieved from
    http://en.wikipedia.org/wiki/File:Abraham_Lincoln_head_o
    n_shoulders_photo_portrait.jpg on March 28, 2009.

12. Editor/Staff, quotes Abraham Lincoln, QuoteDB,
    QuoteDB.com. Retrieved from http://www.quotedb.com.
    quotes/1997 on February 24, 2009.

13. Bharat Rathode, "Teaching Intelligent Design violates First
    Amendment, says Federal Judge," The Earth Times,
    EARTHtimes.org, December 22, 2005. Retrieved from
    http://www.earthtimes.org/articles/show/4757.html on
    February 14, 2008.

14. Editor, "Roger Baldwin," DiscoverTheNetworks.Org, A Guide
    To The Political Left. Retrieved from http://www.
    discoverthenetworks.org/individualProfile.asp?indid-1579
    on October 14, 2008.

15. Staff/Editor, "Norman Thomas," quoted by PatriotsAgainst
    Socialism & Totalarism. Retrieved from
    http://patriotsagainstsocialism.org/ on December 23, 2008.

16. Staff/Editor, "Nikita Khrushchev," Nikita Khrushchev
    Quote/Quotations, Liberty-Tree.ca. Retrieved from
    http://quotes.liberty-tree.ca/quote/nikita_khrushchev_
    quote_7174 on February 14, 2009.

Chapter 9—Freedom Of Thought VS. What's Not Being Taught

1. Sean D. Pitman (compiled by), "Thoughts on Evolution From Scientists and Other Intellectuals," (quotes Wolfgang Smith (1988) Teilhardism and the New Religion: A Thorough Analysis of The Teachings of Pierre Teilhard de Chardin, Rockford, Illinois: Tan Books & Publishers Inc., p. 2), p. 12. Retrieved from http://www.detectingdesign.com/quotesfromscientists.html on September 26, 2008 which was retrieved from original website http://naturalselection.0catch.com/Files/quotesfromscientists.html.

2. Editor/Staff, Church of God, International, "What are the Lies of Evolution?"—Part 2, p. 9. (quotes George Kocan, "Evolution Isn't Faith But Theory," Chicago Tribune, Monday, April 21, 1980.) Retrieved from http://www.biblestudy.org/basicart/what-are-the-lies-of-evolution-2.html on February 14, 2009. See also: Editor/Compiler, Quotable Quotes: "What Are Scientists And others Really saying About Evolution?"—p. 14, (www.2Christ.org), Retrieved from http://www.2christ.org/quotes/ on April 11, 2009.

3. Louis P. Sheldon, "Education Secretary Attacked for Urging Christian Values in Schools," Traditional Values Coalition. Retrieved from http://www.traditionalvalues.org/modules.php?sid=861 on October 16, 2008.

4. Donald E. Wildmon, "Bush Administration Urges "Pledge Across America,," AFA Online, American Family Association, AFA Activism Action Alert, October 11, 2001. Retrieved from http://www.afa.net/activism/aa101101.asp on October 16, 2008.

5. Staff/Editor, "ACLU—One Nation Under God reports ACLU activity that threatens our religious liberty by engaging in religious cleansing of American Society," One Nation Under God website, Retrieved from http://onenationundergod.org/wl_aclu.html on October 15, 2008.

6. Ann Coulter comments, Darwin's Deadly Legacy, Jerry Newcombe and John Rabe, Coral Ridge Ministries, DVD, 2006 (www.CoralRidge.org). (Coulter refers to Richard Weikart's 2004 book, From Darwin to Hitler: Evolutionary Ethics, Eugenics, and Racism in Germany.)

7. Richard Weikart, <u>Praise for *From Darwin to Hitler*</u>, Nancy Pearcey, <u>Review of Richard Weikart's book</u>: *From Darwin to Hitler: Evolutionary Ethics, Eugenics, and Racism in Germany* (Palgrave Macmillan, New York, 2004 (paperback edition in 2005), <u>From Darwin to Hitler website</u>. Retrieved from http://www.csustan.edu/History/Faculty/Weikart/FromDa rwintoHitler.htm on October 15, 2008.

8. H.S. Lipson, "A Physicist Looks at Evolution," Physics Bulletin, Vol. 31, p. 138 (1980), http://www.pathlights.com/ce_encyclopedia/Encyclopedia/ 21soc04.htm.

9. D. James Kennedy quotes Eric Harris, <u>Darwin's Deadly Legacy</u>, DVD, Jerry Newcombe and John Rabe, D. James Kennedy (Host), Coral Ridge Ministries, 2006 (www.CoralRidge.org).

10. Ibid 9. (Quotes Eric Harris, <u>Darwin's Deadly Legacy</u>, DVD, Jerry Newcombe and John Rabe, D. James Kennedy (Host), Coral Ridge Ministries, 2006 (www.CoralRidge.org).

11. Richard Dawkins, "Put Your Money on Evolution," *The New York Times* (April 9, 1989) section VII, p. 35. Retrieved from http://www.mercatornet.com/articles/view/evolution_speci al_issuedarwinism_strikes_back/, January 17, 2009. (Additional source: Richard Dawkins: byline/review of book: *Blueprints: Solving the Mystery of Evolution* by Maitland A. Edey and Donald C. Johanson, *The New York Times*, April 9, 1989. Retrieved January 15, 2009, from http://www.simonyi.ox.ac.uk/dawkins/WorldOfDawkins-archive/Dawkins/Work/Reviews/1989–04–09review_blueprint.shtml; Comment was also then quoted from Josh Gilder, a creationist, in his critical review, "PBS's 'Evolution' series is propaganda, not science" (September, 2001). Additional source: D. James Kennedy (Host), <u>Darwin's Deadly Legacy</u>, Jerry Newcombe and John Rabe, DVD, Coral Ridge Ministries Media, Inc., 2006–2007 (www.CoralRidge.org).

12. Marvin Olasky, (Phillip E. Johnson quote), <u>Roots of genocide</u>, Interview: Richard Weikart on how Hitler was Darwin's ideological grandson by Marvin Olasky, WORLD Magazine, Vol. 20, No. 16, April 23, 2005. Retrieved from http://www.worldmag.com/ printer.cfm?id=10552 on October 14, 2008.

13. Staff/Editor, "Explore Roloff Farm," <u>Little People, Big World</u>, The Learning Channel Fansites (TLC), Retrieved from http://tlc.discovery.com/fansites/lpbw/roloff_farm/explore. html on October 19, 2008. Additional source: "Little People,

Big World." From Wikipedia—the free encyclopedia. Retrieved from http://en.wikipedia.org/wiki/Little_People_ Big_World on October 19, 2008.

14. George Foreman, III, George Foreman, III comments to author, Houston, TX, May, 2008.

Chapter 10—Media Matters

1. Jody Brown and Ed Vitagliano, "ABC's Latest 'Jesus' Special Denigrates Christianity, Groups Say," American Family Association (www.afa.net), AgapePress news, p. 1, April 6, 2004. Retrieved from http://headlines.agapepress.org/ archive/4/afa/62004f.asp on October 19, 2008.

2. Ibid.

3. Patrick J. Mahoney, WDC Guest Writer, "Protests Demands ESPN Releases Tape of Anchor's Profane Comments," WDC Media News, Christian News and media Agency, January 1, 2008, p 1. Retrieved from http://www.wdcmedia.com/ newsArticle.php?ID=3780 on October 19, 2008.

4. Jody Brown and Ed Vitagliano, "ABC's Latest 'Jesus' Special Denigrates Christianity, Groups Say," American Family Association, AgapePress news, p. 2, April 6, 2004. Comments on Bernard Goldberg regarding his best-selling 2001 book: *Bias: A CBS Insider Exposes How the Media Distort the News*. Retrieved from http://headlines.agapepress.org/archive/4/afa/62004f.asp on October 19, 2008.

5. Editor/Staff, NationalGeographic.com, "New Birdlike Dinosaurs From China Are True Missing Link," T. Rex With Feathers? @ nationalgeographic.com, Press Events (October 15,1999), p. 1—referring to upcoming November 1999 issue of National Geographic magazine. Retrieved from http://www.nationalgeographic.com/events/99/feather/index.html on December 10, 2008.

6. "Evolution myths." CreationWiki, the encyclopedia of creation science. "Archaeoraptor, p. 2. Source: Christopher P. Sloan, "Feathers For T-Rex?," National Geographic Magazine, Vol. 196, No. 5, November, 1999, pp. pp.98–107 (99,100,105). 99,100,105.

7. Storrs L. Olson, "Open Letter TO: ...National Geographic Society," from Storrs L. Olson, National Museum of Natural History, Smithsonian Institution, Washington, DC, November 1, 1999.

8. Charles Darwin, Introduction to *On The Origin of Species*, John Murray Publisher, November 24, 1859.

9. William J. Federer comments, <u>One Nation Under God</u>, Jerry Newcombe and John Rabe, Coral Ridge Ministries, DVD, 2005, hosted by Dr. D. James Kennedy (www.CoralRidge.org).
10. E. Yaroslavsky, Landmarks in the Life of Stalin (Moscow: Foreign Languages Publishing House, 1940), pp. 8–12. Source: Creation Tips, "Why evolution creates tyrants like Hitler, Stalin and Trotsky," Staff article. Retrieved from http://www.users.onaustralia.com.au/rdoolan/tyrants.html; on July 11, 2008.

Chapter 11—Part III: Tomorrow—Cloning & Genetic Engineering

1. Editor/Staff, <u>MedicineNet.com</u>, Medicine.Net Doctors, "Definition of Cloning," Cloning definition—Medical Dictionary definitions of popular medical terms easily defined. Retrieved from http://www.medterms.com/script/main/art.asp?articlekey= 2756 on October 28, 2008. (See also: MedicineNet.com Doctors, *Webster's New World™ Medical Dictionary*, 3rd Edition, Wiley Publishing, Inc., Hoboken, NJ, May, 2008.)
2. Psalm 139:13–16, "Scripture taken from the NEW AMERICAN STANDARD BIBLE®, Copyright © 1960, 1962, 1963, 1968, 1971, 1971, 1972, 1975, 1977, 1995 by The Lockman Foundation. Used by permission."
3. <u>Life Issues Institute</u>, "April 10, 2002 White House address on cloning by President George W. Bush." Retrieved from http://www.lifeissues.org/cloningstemcell/bush.html on February 17, 2009.
4. Photo: Charles Darwin depicted as an ape, Hornet Magazine, 1871, The media is lacking author information: http://en.wikipedia.org/wiki/File:Darwin_ape.jpg on January 14, 2009.
5. Fazale (Fuz) Rana, "Do Humans and Chimps Belong In The Same Genus?," Reasons To Believe website (www.reasons.org). Retrieved from http://www.reasons.org/ resources/apologetics/humans_chimps_same_genus.shtml on October 6, 2008
6. Anthony Carpi, "The Cell," in Dr. Carpi's course study on <u>The Natural Science Pages</u>—NSC 107: An introduction to Science in Society at John Jay College of the City University of New York designed for general use and all visitors welcome. Retrieved

fromhttp://web.jjay.cuny.edu/~acarpi/NSC/13-cells.htm on March 30, 2009.

7. John MacArthur, <u>Creation: Believe It or Not</u>, CD, www.gty.com, 1999.

8. ARN Staff, ARN Quote Library—Relevant Quotes, quotes "Cohen, I L., Darwin Was Wrong: A Study in Probabilities," NW Research Publications, Inc., New York, p. 209, Dec. 1984. Retrieved from http://www.arn.org/blogsq/index.php?title=survival_of_the_fittest_and_natural_sele&more=1&c=1&tb=1&pb=1 on September 3, 2008.

9. Gordon Rattray Taylor (former Chief Science Advisor, BBC Television), *The Great Evolution Mystery,* Harper & Row Publishers, Inc., New York, 1983, p. 48.

10. Richard Paley, (compiler), "Quotations Compiled by Dr. Richard Paley," Objective Ministries, ObjectiveMinistries.org, quotes Albert Szent-Gyorgi, Nobel Laureate, Medicine. Retrieved from http://objectministries.org/creation/quotes.html on October 28, 2008.

11. Jonathan Wells comments, *Redeeming Darwin: Discovering the Designer,* DVD, Probe Ministries and EvanTell, 2007.

12. Jonathan Wells comments, <u>Darwin's Deadly Legacy</u>, Jerry Newcombe and John Rabe, Coral Ridge Media Ministries, DVD, 2006.

13. Ian Taylor comments, <u>Darwin's Deadly Legacy</u>, Jerry Newcombe and John Rabe, Coral Ridge Media Ministries, DVD, 2006–2007 (www.CoralRidge.org). (Ian Taylor, *"In the Minds of Men: Darwin and the New World Order,* Tfe Pub; 3rd edition, October 1996.)

14. John Leslie, "Cosmology, Probability, and the Need to Explain Life," in Scientific American and Understanding, pp. 53, 64–65; E.J. Ambrose, *Nature and Origin of the Biological World,*" Ellis Horwood Publisher, Chichester (West Sussex): 1982, p. 135.

15. Alan E. Guttmacher & Francis Collins, *"Realizing the Promise of Genomics in Biological Research,"* JAMA (The Journal of the American Medical Association—Commentary), Vol. 294 (294:1399–1402), No. 11, September 21, 2005. Retrieved from http://jama.amaassn.org/cgi/content/full/294/11/1399?ijkey=OV9zY.Uujjq.Q&keytype=ref&siteid=amajnls on October 29, 2008.

16. Staff/Editor, "Facts on Health Care in the U.S. and Abroad," Heritage Policy Blog, Heritage Foundation (www.heritage.org). Retrieved from

http://www.heritage.org/press/dailybriefing/policyweblog.cf
m?blogid=AB69914D-A0C9-D18A-0F81A5C44F919338 on
April 1, 2009.

Chapter 12—God vs. Darwin

1. Jonathan *Wells, Icons of Evolution: Science or Myth? Why
   Much of What We Teach About Evolution is Wrong*, Regnery
   Publishing, Washington, D.C., 2000, p. 24.
2. Henry Morris, "Evolution, Thermodynamics and Entropy,"
   Institute for Creation Research (www.icr.org), May, 1973,
   quotes Isaac Asimov, "In the Game of Energy and
   Thermodynamics You Can't Break Even," *Smithsonian
   Institute Journal*, June, 1970, p. 6. Retrieved from
   http://www.icr.org/articles/print/51/ on January 11, 2009.
3. Isaiah 1:18—"Scripture taken from the NEW AMERICAN
   STANDARD BIBLE®, Copyright © 1960, 1962, 1963, 1968,
   1971, 1971, 1972, 1975, 1977, 1995 by The Lockman
   Foundation. Used by permission."

Chapter 13—Cheating Science

1. Editor/Staff, FoxNews.Com (SciTech), "Biologist: I Lost My
   Job Because I Don't Believe in Evolution," source:
   Associated Press, Boston, MA, December 11, 2007. Retrieved
   from
   http://www.foxnews.com/story/0,2933,316406,00.html on
   January 22, 2008.
2. Ibid.
3. Barbara Weller, Attorney Barbara Weller comments to
   author, April 1, 2009.
4. Royal Truman, "What biology textbooks never told you about
   evolution—A review of *Icons of Evolution: Science or Myth?
   Why Much of What We Teach About Evolution is Wrong* by
   Jonathan Wells, Regnery Publishing, Inc. Washington, DC,
   2000," AnswersInGenesis, www.answersingenesis.org, First
   published: TJ 15(2):17–24, August, 2001. Retrieved from
   http://www.answersingenesis.org/tj/v15/i2/textbooks.asp
   on October 30, 2008.
5. "Luke." From Conservapedia, www.conservapedia.com,
   quotes Sir William Ramsay, *The Bearing of Recent Discovery
   on the Trustworthiness of the New Testament* (reprinted from
   the 1915 edition), Baker House Books, Grand Rapids, MI,
   1953, p. 222.

6. "Luke." From <u>Conservapedia</u>, www.conservapedia.com, quotes Sir William Ramsay, *St. Paul the Traveller and the Roman Citizen* (reprinted from the 1897 edition), Baker House Books, Grand Rapids, MI, 1962, p. 81.

7. Editor/Staff, <u>WORDsearchBible.com</u>, WORDsearch Corp., Austin, TX: synopsis on book by Sir William Mitchell Ramsey, *Luke the Physician*, Hodder & Stoughton, London, 1908. Retrieved from http://www.wordsearchbible.com/catalog/product.php?pid =1719 on July 20, 2008.

8. Lenel Parish, Whitman College, <u>Whitman Receives First Edition "Origin of Species,"</u> Whitman College News Service, News Release, September 16, 2004. Retrieved from http://www.whitman.edu/content/news/Darwins on July 25, 2008.

9. Kerby Anderson, Redeeming Darwin: Discovering the Designer, DVD, Probe Ministries and EvanTell, 2007.

10. <u>ThinkExist.com Quotations</u>, "Albert Einstein quotes," Contributors, www.thinkexist.com. "Albert Einstein quotes," Contributors, <u>ThinkExist.com Quotations</u>, www.thinkexist.com. Retrieved from http://thinkexist.com/quotation/i_want_to_know_how_god_ created_this_world-i_am/15496.html on November 5, 2008.

11. Editor/Contributors, "Sickle Cell Anemia, Who Is At Risk," <u>National Heart, Lung and Blood Institute Diseases and Conditions Index</u>, www.nhlbi.nih.gov, U.S. Department of Health & Human Services—National Institutes of Health. Retrieved from http://www.nhlbi.nih.gov/health/dci/Diseases/Sca/SCA_W hoIsAtRisk.html on February 19, 2009.

12. Stephen J. Gould, *The Mismeasure of Man*, Norton & Company, Inc., New York, NY, 1981, p. 323.

Chapter 14—Social Darwinism Tomorrow

1. Denyse O'Leary, "Darwin and scientific racism," The I.D. Report—Part 2, (www.arn.org), February 20, 2008. Retrieved from http://www.arn.org/blogs/index.php/2/2008/02/20/part_t wo_darwin_and_scientific_racism on December 10, 2008.

2. Denyse O'Leary, "Darwin and the Holocaust," The I.D. Report—Part 3, (www.arn.org) February 20, 2008. Retrieved from http://www.arn.org/blogs/index.php/2/2008/02/20/part_t wo_darwin_and_scientific_racism on December 10, 2008. .

3. Gary Butner, ed., Quotes Evolution—Quotes Regarding Evolution/Creation/ID (www.errantskeptics.org), quotes Dr. T. N. Tahmisian, U.S. Atomic Energy Commission physiologist, in <u>The Fresno Bee</u>, p. 1-B, August 20, 1959 as quoted by N. J. Mitchell, *Evolution and the Emperor's New Clothes*, title page, Roydon Publications, UK, 1983. Retrieved from http://www.errantskeptics.org/Quotes_Regarding_Creation_Evolution.htm on November 5, 2008.

4. Richard M. Riss, *Christian Evidences*, "Secular Humanism: An Evaluation (www.grmi.com)," quotes Malcolm Muggeridge, *The End of Christendom, But Not of Christ*, (William B. Eerdmans Publishing Co., Grand Rapids, MI, 1980, pp. 4–5. Retrieved from http://www.grmi.org/Richard_Riss/evidences2/05sec.html on November 5, 2008.

5. Jone Johnson Lewis (compiler), "William Golding," <u>Wisdom Quotes</u> (www.wisonmquotes.com), Original source: William G. Golding, "Belief and Creativity," Lecture, Hamburg, Germany, April 11, 1980. Retrieved from http://www.wisdomquotes.com/001500.html on February 18, 2009.

6. Richard Dawkins, "Put Your Money on Evolution," *The New York Times* (April 9, 1989) section VII, p. 35. Retrieved from http://www.mercatornet.com/articles/view/evolution_special_issuedarwinism_strikes_back/, January 17, 2009. (Additional source: Richard Dawkins: byline/review of book: *Blueprints: Solving the Mystery of Evolution* by Maitland A. Edey and Donald C. Johanson, *The New York Times*, April 9, 1989. Retrieved January 15, 2009, from http://www.simonyi.ox.ac.uk/dawkins/WorldOfDawkins-archive/Dawkins/Work/Reviews/1989–04–09review_blueprint.shtml; Comment was also then quoted from Josh Gilder, a creationist, in his critical review, "PBS's 'Evolution' series is propaganda, not science" (September, 2001). Additional source: D. James Kennedy (Host), <u>Darwin's Deadly Legacy</u>, Jerry Newcombe and John Rabe, DVD, Coral Ridge Ministries Media, Inc., 2006–2007 (www.CoralRidge.org).

7. Bryan Bissell, "Great Scientists Who Believe," SDA Global (www.english.sdaglobal.org), p. 5, quotes Dr. Arthur E. Wilder-Smith in Willem J.J. Glashouwer and Paul S. Taylor, *The Origin of the Universe*, (P.O. Box 200, Gilbert, AZ 85299 USA: Eden Communications and Standard Media, 1983). Retrieved from

http://english.sdaglobal.org/research/sctstbel.htm on
September 4, 2008.

8. Stephen Jay Gould, Professor of Geology and Paleontology
(taught biology, geology and history of science), Harvard
University–deceased May, 2002), "Is A New and General
Theory of Evolution Emerging?"—Paleobiology, Vol. 6, No. 1,
January 1980, pp. 119–130.

9. Editor/Staff, ThinkExist.com, Albert Einstein quotation,
Retrieved from
http://thinkexist.com/quotation/i_want_to_know_how_god_
created_this_world-i_am/15496.html on October 30, 2008.

10. Bryan Bissell, "Great Scientists Who Believe," Relativity
Theory, p. 35, SDA Global (www.english.sdaglobal.org).
Retrieved from
http://english.sdaglobal.org/research/sctstbel.htm#_Toc71
912181 on September 5, 2008.

11. Galileo Galilei. Wikiquote. Retrieved from
http://ang.wikiquote.org/wiki/Galileo_Galilei on April 6,
2009.

12. "Isaac Newton." From Wikipedia, the free encyclopedia.
Source: J.H. Tiner, *Isaac Newton: Inventor, Scientist and
Teacher*. Milford, Michigan, U.S.: Mott Media, 1975.
Retrieved from http://en.wikipedia.org/wiki/Isaac_Newton
on March 13, 2009. (Also, provides the source material on
Newton's life and credentials.)

13. Vance Ferrell, *SCIENCE VS. EVOLUTION*
(www.pathlights.com), Chapter 23a: "Scientists Speak" (from
the Evolution Disproved Series), quotes W.R. Thompson,
Introduction to Charles Darwin, *Origin of Species*, 1956
edition; W.R. Thompson, "Introduction," Origin of Species;
statement reprinted in *Journal of the American Affiliation*,
March, 1960. Retrieved from
http://www.pathlights.com/ce_encyclopedia/sci-
ev/sci_vs_ev_23.htm on September 22, 2007.

14. Charles Darwin, "The Life and Letters of Charles Darwin,
p.294. See Also: Charles Darwin from Conservapedia.
Retrieved from http://www.conservapedia.com/Charles_
Darwin on April 6, 2009—(Dr. Walt Brown, Center for
Scientific Creationism, http://www.creationscience.com/
onlinebook/ReferencesandNotes12.html).

15. Vance Ferrell, *SCIENCE VS. EVOLUTION*, Chapter 23a:
Scientists Speak, *I.L. Cohen, Darwin Was Wrong: A Study in
Probabilities. (1985). Retrieved from
http://www.pathlights.com/ce_encyclopedia/sci-
ev/sci_vs_ev_23.htm on September 26, 2008. Source:*

Cohen, I.L. *Darwin Was Wrong: A Study in Probabilities*, New York: NW Research Publications, Inc., 1984, pp. 214–215. Retrieved from http://www.goodschools.com/darwin.htm on November 5, 2008.

16. Fellowship Contributors, See: www.bible411.com, Bible Students Congregation of New Brunswick, quotes Robert Jastrow, *God and the Astronomers*, W.W. Norton & Company, New York, July 1978, p. 116. Retrieved from http://www.bible411.com/creation/chapter2.htm on February 21, 2009.

Chapter 15—Rights & Freedoms Now

1. William Dembski, quote from Jonathan Wells, "Wells vs. Shermer at Cato Institute" (October 14, 2006), Uncommon Descent: Serving the Intelligent Design Community. Retrieved from http://www.uncommondescent.com/evolution/wells-vs-shermer-at-cato-institute/ on October 7, 2008.

2. Colin Patterson, Senior Paleontologist, British Museum of Natural History, London Keynote address at the American Museum of Natural History, New York City, November 5, 1981—quoted by Gary Butner, Errant Skeptics Research Institute (www.errantskeptics.org). Retrieved from http://www.errantskeptics.org/Quotes_Regarding_Creation_Evolution.htm on November 5, 2008.

3. Staff Report, United States House of Representatives Committee On Government Reform (December, 2006), "Intolerance And The Politicization Of Science At The Smithsonian: Smithsonian's Top Officials Permit The Demotion And Harassment Of Scientist Skeptical Of Darwinian Evolution," Executive Summary, December 11, 2006, p. 3. Retrieved from http://www.souder.house.gov/_files/IntoleranceandthePoliti cizationofScienceattheSmithsonian.pdf on March 4, 2009.

4. Staff Report, United States House of Representatives Committee On Government Reform (December, 2006), "Intolerance And The Politicization Of Science At The Smithsonian: Smithsonian's Top Officials Permit The Demotion And Harassment Of Scientist Skeptical Of Darwinian Evolution," Executive Summary, December 11, 2006, p. 3–4. Retrieved from http://www.souder.house.gov/_files/IntoleranceandthePoliti cizationofScienceattheSmithsonian.pdf on March 4, 2009.

5. Ibid, p. 5. ("...statements," March 22, 2005, 9:53 AM, email to ... with attached memo dated February 8, 2005.)
6. Ibid, p. 5. (Reference Investigative subcommittee staff report: ..."Re: misc," February 22, 2005, 9:38 AM, email to ....)
7. Ibid, p. 18.
8. David Klinghoffer, The Branding Of A Heretic: *Are Religious Scientists Unwelcome At The Smithsonian?*," The Wall Street Journal, January 28, 2005.
9. Hugh Ross, About US: "Science and Faith: Allies and Enemies," Reasons To Believe, (www.reasons.org). Retrieved from http://www.reasons.org/about/index.shtml on October 10,2008.
10. Bryan Bissell, Great Scientists Who Believe (Part 2), SDA Global (www.english.sdaglobal.org), quotes Dr. J.Y. Chen, Chinese Paleontologist, p. 4. Retrieved from http://english.sdaglobal.org/research/sctstbel.htm on September 4, 2008.
11. I.L. Cohen, *Darwin Was Wrong—A Study in Probability*, New Research Publications, Inc., Greenvale, New York, 1984, pp. 209–210. Source: Good Schools, "Darwin's Theory of Evolution—A Notion Rooted Deep in Racism, but not in Science!" (www.goodschools.com/darwin.htm) Retrieved from http://cc.msnscache.com/cache.aspx?q=%22www+goodsch ools+com+darwin+htm%22&d=75767307643843&mkt=en-US&setlang=en-US&w=bbd2a4cc,989259c2 on April 12, 2009. Additional source: Editor/Compiler, Quotable Quotes (p. 11): "What Are Scientists and Others Really Saying About Evolution?" (www.2Christ.org), Retrieved from http://www.2christ.org/quotes/ on April 11, 2009.
12. Richard Lewontin, *"Billions and Billions of Demons,"* New York Review of Books, January 9, 1997, p. 28–31. See full quote at: http://www.veritas-ucsb.org/library/ origins/quotes/ mutations.html accessed on March 30, 2009. See also: Richard Lewontin, quoted by Phillip E. Johnson in the unraveling of scientific materialism, *Objections Sustained: Subversive Essays on Evolution, Law and Culture*, InterVarsity press, Downers Grove, IL, April, 2000, p.71–72.
13. Neal C. Gillespie, *Charles Darwin and the Problem of Creation*, University of Chicago Press (Tx), June, 1977, p.2. Quotes Charles Darwin from a letter to Harvard biology professor Asa Gray.

Additional copies of
DARWIN'S RACISTS
Yesterday, Today and Tomorrow
are available at www.virtualbookworm.com or
www.DarwinsRacists.com.

CPSIA information can be obtained at www.ICGtesting.com
Printed in the USA
LVOW010035221011

251461LV00004B/1/P